Priceless Knowledge?

Priceless Knowledge?

Natural Science in Economic Perspective

NICHOLAS RESCHER

ROWMAN & LITTLEFIELD PUBLISHERS, INC.
Lanham • Boulder • New York • London

ROWMAN & LITTLEFIELD PUBLISHERS, INC.

Published in the United States of America
by Rowman & Littlefield Publishers, Inc.
4720 Boston Way, Lanham, Maryland 20706

3 Henrietta Street
London WC2E 8LU, England

British Cataloging in Publication Information Available

Library of Congress Cataloging-in-Publication Data

Rescher, Nicholas
Priceless knowledge? : natural science in economic perspective / Nicholas Rescher.
p. cm.
Includes bibliographical references and index.
1. Research—Economic aspects. 2. Natural history—Research—Economic aspects.
I. Title
Q180.55.E25R45 1996 338.4'75072—dc20 96-18508 CIP

ISBN 0–8476–8244–7 (cloth : alk. paper)
ISBN 0–8476–8245–5 (pbk. : alk. paper)

Printed in the United States of America

♾™ The paper used in this publication meets the minimum requirements of American
National Standard for Information Sciences—Permanence of Paper for Printed Library
Materials, ANSI Z39.48–1984.

Contents

PREFACE

This book brings together in a connected and (so I hope) coherent fashion the substance and ideas and arguments that I have put forward on various occasions over the past twenty years. It seeks to give a unified expression to an economically realistic view of the future prospects of natural science whose various aspects I have treated in different publications—a view that gives pronounced emphasis to the theme of limits and limitations.

I am grateful to Professors Wenceslao González and Michele Marsonet for their encouragement of this project, and I want to thank Estelle Burris for her help in producing an elegantly word-processed text suitable for the printer's use.

Nicholas Rescher
Pittsburgh, PA
September 1995

INTRODUCTION

The line of thought developed in this book, while insofar as correct of far-reaching significance, is very straightforward. It maintains that although natural science is by nature an incompletable project, and a theoretical prospect of further progress lies ever before us, nevertheless, the ongoing pursuit of this prospect is subject to a drastic increase in cost. The basic idea is (1) that—contrary to what many people nowadays apparently believe—it is quite mistaken to think that science is coming to an end with all the really big discoveries pretty much in hand; (2) that ongoing progress in science, while theoretically available, is nevertheless something that demands an ever more sophisticated technology for the acquisition and management of data; and (3) that the development of such an ever more powerful technology requires the commitment of ever greater resources—and thereby calls for meeting ever greater costs. The fact is that scientific inquiry as a human enterprise involves an indefinitely extensible (and so, ultimately imperfectable) technology of inquiry in a way that renders the progress of the enterprise on the *cognitive* side dependent upon the investment of resources (by way of money, time, talent, and effort) on the *material* side.

A central element of the analysis lies in the operation of a law of logarithmic returns relating the growth of knowledge to the acquisition of mere information. Its operation pivots on two ideas: (1) that the acquisition of new *information* in science comes at a rate that is roughly proportional to resource expenditures, but (2) that new *knowledge* stands in proportion not with the actual volume of new information but only with its logarithm. This of course means that a uniform, constant increase in the growth of knowledge requires an exponential increase in resource investment. Scientific knowledge may be priceless in value but it is emphatically not priceless in point of cost.

Only by continually committing massively increasing resources to scientific work could a steady pace of progress be maintained. Since this is economically infeasible in practice, there looms before us an unavoidable and drastic deceleration in the pace of scientific progress. The theoretical prospect of unendingly smooth progress in science encounters an insuperable practical obstacle to its realization—a process of cost inflation where innovations of the same inherent magnitude become ever more expensive. In a world of finite resources the implications of this circumstance are only too obvious: scientific discovery grows increasingly expensive, and we are involved in a situation of diminishing returns on resource investment that will, in due course, engender a drastic slowing in the pace of progress.

This line of thought leads to conclusions of a distinctly sobering tenor. It means that while the inherent potential of discovery in science is theoretically unending, we actually face the reality of limits. The idea of perfecting science is accordingly impracticable—not on abstractly theoretical but on strictly economic grounds. This state of affairs has obvious implications for the imperfectability of our knowledge of the world. It means that our cognitive probes cannot plumb nature to the very bottom of its depth. There are no *theoretical* limits to natural science—no "end of the road" that we are managing to reach or even approach. All the same, inexorable limits and limitations are rooted in the fact that there is only so much that we can afford to do. Natural science, like any other constructive venture on which we humans can embark, is subject to suboptimality-constraining economic limitations.

This, then, is the story line that unfolds in the present deliberations. It points toward a somber and sobering recognition of the cognitive limitations that are inherent in the human condition, confronting us with a Sisyphus-reminiscent task in our efforts to advance natural science in the endeavor to establish cognitive mastery over the world we live in. Natural science is thus imperfectable on practical—and ultimately economic—grounds, and this means that our scientific grasp of the natural world will ever remain substantially incomplete.

It follows the thesis of scientific realism that science correctly and comprehensively describes the real world: that science is, or ever will be, free from significant errors of commission and omission is no more than a pipe dream. The economic realities of the conditions under which our scientific inquiries into nature do (and must!) proceed are such that our knowledge is imperfectable and that the putative "laws of nature" that our science provides us are no better than best estimates whose almost certain imperfections we have not as yet discovered.[1]

NOTES

1. Here the present analysis harmonizes with the thesis of Nancy Cartwright's provocative book *How the Laws of Nature Lie* (Oxford: Clarendon Press, 1983). To be sure, Cartwright argues for this position in a totally different way, maintaining that "There are no rigorous solutions for real-life problems [in physics]. Approximation and adjustments are required whenever theory treats reality" (p. 13).

Chapter One

THE ECONOMIC DIMENSION
OF RATIONAL INQUIRY

SYNOPSIS

(1) The development of knowledge is a human activity that, as such, has an unavoidably economic dimension. It exacts a considerable price from us, and it brings great benefits as well. (2) This state of affairs means that economic cost-benefit considerations will also have to play a crucial role in the cognitive domain.

1. COGNITION AS AN ECONOMIC ACTIVITY

The changeable nature of human knowledge has been recognized since classical antiquity.[1] The history of science and of inquiry in all its forms clearly shows that facts, theories, concepts, methods are not endowed with a fixed and stable nature but are processual entities that reflect the ever-changing cognitive state of the art of a place, time, and cultural modus operandi. Science is not an object of some sort, a thinglike "body of knowledge," it is an activity—a dynamic cognitive enterprise geared to the creation and active manipulation of information. Human cognition is a process that actively develops, and with this development we have not only change but the emergence of a novelty that radically divides the present from what has gone before. Our cognitive endeavors are involved in a constant process of transformation. Since the rise of science, no two human generations have viewed the world and its contents in just the same way. Science, properly understood, is not a body of thesis or theories, but a process—an ongoing project of inquiry whose products are ever changing.

The development of knowledge—and of science in particular—is thus a human activity that, like any other, involves the expenditure of time, effort, and other resources—and accordingly has an ineliminable economic dimension. Knowledge possesses an ineliminable economic dimension because of its nature as an activity that, as such, has a substantial involvement with costs and benefits. Virtually every aspect of the way we acquire, maintain, and use our knowledge can be properly understood and explained only from an economic point of view. Throughout the entire range of our endeavors in this world, we are involved in the expenditure of limited resources, and knowledge is no exception to this rule. Its acquisition, processing, storage, retrieval, and utilization are activities that, like any other human endeavor, engender costs. Over and above this practical dimension there are also certain purely cognitive disabilities and negativities— that it, costs—involved in the lack of knowledge, in ignorance, error, and confusion. On the positive side, it has come to be increasingly apparent in recent years that knowledge is *cognitive capital* and that its extraction and consolidation involve the creation of intellectual assets in which both producers and users have a very real interest. Any theory of knowledge that ignores this economic aspect of the matter does so to the detriment of its own adequacy.

That greatest of American philosophers, Charles Sanders Peirce (1839-1914) proposed to construe the "economy of research" at issue in knowledge development in terms of the sort of balance of assets and liabilities that we today would call cost-benefit analysis.[2] Peirce insisted that one must recognize the inevitably economic nature of any human enterprise —inquiry included:

> The value of knowledge is, for the purpose of science, in one sense absolute. It is not to be measured, it may be said, in money; in one sense that is true. But knowledge that leads to other knowledge is more valuable in proportion to the trouble it saves in the way of expenditure to get that other knowledge. Having a certain fund of energy, time, money, etc., all of which are merchantable articles to spend upon research, the question is how much is to be allowed to each investigation; and *for* us the value of that investigation is the amount of money it will pay us to spend upon it. *Relatively*, therefore, knowledge, even of a purely scientific kind, has a money value. (*Collected Papers*, vol. 1 (Cambridge, Mass.: Harvard University Press,1931), sect. 1.122 [ca. 1896])

On the side of benefits of scientific claims, Peirce also recognized a wide variety of epistemic factors: closeness of fit to data, explanatory value, novelty, simplicity, accuracy of detail, precision, parsimony, concordance with other accepted theories, even antecedent likelihood and intuitive appeal. He placed in the liability column those cost-geared factors of "the dismal science": the expenditure of time, effort, energy, and money needed to secure and substantiate our claims. This view of the matter is entirely appropriate, though, to be sure, the introduction of such an economic perspective does not, of course, detract from the value of the quest for knowledge as an intrinsically worthy venture endowed with a perfectly valid *l'art pour l'art* aspect.

Philosophical epistemologists subsequent to Peirce have paid regrettably little attention to these matters.[3] Indeed, they often proceed on the tacit assumption that information is something that is economically costless—a free good that comes to rational inquirers without expenditure and effort of course. Even casual consideration, however, shows that such a view is totally erroneous and unrealistic.

For sure, knowledge brings great benefits. The relief of ignorance is foremost among them. Man has evolved within nature into the ecological niche of an intelligent being. In consequence, the need for understanding, for "knowing one's way about," is one of the most fundamental demands of the human condition. Man is *Homo quaerens*. The need for knowledge is part and parcel to our nature. A deep-rooted demand for information and understanding presses in upon us, and we have little choice but to satisfy it. Once the ball is set rolling, it keeps on under its own momentum—far beyond the limits of strictly practical necessity. The great Norwegian polar explorer Fridtjof Nansen put it well. What drives men to the polar regions, he said, is

> the power to the unknown over the human spirit. As ideas have cleared with the ages, so has this power extended its might, and driven Man willy-nilly onwards along the path of progress. It drives us in to

Nature's hidden powers and secrets, down to the immeasurably little world of the microscopic, and out into the unprobed expanses of the Universe. . . . it gives us no peace until we know this planet on which we live, from the greatest depth of the ocean to the highest layers of the atmosphere. This Power runs like a strand through the whole history of polar exploration. In spite of all declarations of possible profit in one way or another, it was that which, in our hearts, has always driven us back there again, despite all setbacks and suffering.[4]

The discomfort of unknowing is a natural component of human sensibility. To be ignorant of what goes on about us is almost physically painful for us—no doubt because it is so dangerous from an evolutionary point of view. It is a situational imperative for us humans to acquire information about the world. We have questions and we need answers. *Homo sapiens* is a creature who must, by his very nature, feel cognitively at home in the world. The requirement for information, for cognitive orientation within our environment, is as pressing a human need as that for food itself. We are rational animals and must feed our minds even as we must feed our bodies. Relief from ignorance, puzzlement, and cognitive dissonance is one of cognition's most important benefits. These benefits are both positive (pleasures of understanding) and negative (reducing intellectual discomfort through the removal of unknowing and ignorance and the diminution of cognitive dissonance).

The basic human urge to make sense of things is a characteristic aspect of our makeup—we cannot live a satisfactory life in an environment we do not under-

stand. For us intelligent creatures, cognitive orientation is itself a practical need: cognitive disorientation
is physically stressful and distressing. As William
James observed: "It is of the utmost practical importance to an animal that he should have prevision of
the qualities of the objects that surround him."[5]

2. COST EFFECTIVENESS AS A
SALIENT ASPECT OF RATIONALITY

The benefits of knowledge are twofold: theoretical
(or purely cognitive) and practical (or applied). The
theoretical/cognitive benefits of knowledge relate to its
satisfactions in and for itself, for understanding is
indeed an end unto itself and, as such, is the bearer of
important and substantial benefits—benefits that are
purely cognitive, relating to the informativeness of
knowledge as such. The practical benefits of knowledge, on the other hand, relate to its role in guiding
the processes by which we satisfy our (noncognitive)
needs and wants. The satisfaction of our needs for
food, shelter, protection against the elements, and se-
curity against natural and human hazards requires
information, and the satisfaction of mere desiderata
comes into it as well. We can, do, and must put knowledge to work to facilitate the attainment of our goals,
guiding our actions and activities in this world into
productive and rewarding lines. This is where the
practical payoff of knowledge come into play.

The costs (and benefits) of knowledge acquisition will
of course vary with people's conditions and circumstances. Time is of the essence here. The medical
knowledge of the twentieth century was not available

to patients in the eighteenth century—"not for all the tea in China." In pursuing information, as in pursuing food, we have to settle for the best we can obtain at the time. We have questions and need answers—the best answers we can get here and now, regardless of their imperfections. We cannot wait until all returns are in. Our needs and wants impel us to resolve our questions by means of the best available answers we can get. What matters for us is not ideal and certain knowledge in the light of complete and perfected information, but getting the best estimate that is actually obtainable here and now.

The impetus to inquiry—to investigation, research, and acquisition of information about the world we live in—can accordingly be validated in strictly economic terms with a view to potential costs and benefits of both theoretical and practical sorts. We humans need to achieve both an intellectual and a physical accommodation to our environment.

The ancients saw man as the rational animal (*zoôn logon echôn*), set apart from other creatures by capacities for speech and deliberation. Under the precedent of Greek philosophy, Western thinkers have generally deemed the use of thought for the guidance of our actions to be at once the glory and the duty of *Homo sapiens*.

To behave rationally is to make use of one's intelligence to figure out the best thing to do in the circumstances. Rationality is a matter of the intelligent pursuit of appropriate objectives; it consists in the use of reason to resolve choices in the best feasible way. Above all, it calls for the intelligent pursuit of appropriate ends, for the effective and efficient cultivation of appropriately appreciated benefits. Rationality requires doing the best one can with the means at one's disposal, striving for the best results that one can expect to achieve within the range of one's resources, specifically including one's intellectual resources. Be it in matters of belief, action, or evaluation, its mission centers about the deliberate endeavor to secure an optimally favorable balance of benefits relative to expenditure.

Accordingly, rationality has an ineliminable economic dimension. The optimal use of resources is, after all, a crucial aspect of rationality. It is against reason to expend more resources on the realization of a given end than one needs to.[6] It is against reason also to expend more resources on the pursuit of a goal than it is worth—to do things in a more complex, inefficient, or ineffective way than is necessary in the circumstances. It is also against reason, though, to expend fewer resources in the pursuit of a goal than it is worth, unless these resources can be used to even better effect elsewhere. Cost effectiveness—the proper coordination of costs and benefits in the pursuit of our ends—is an indispensable requisite of rationality.

This general situation obtains with particular force where the transaction of our specifically *cognitive* business is concerned. With any source of information or method of information acquisition, two salient questions arise:

1. *Utility*: How useful is it; how often do we have occasion/need to make use of it; how significant are the issues that rest on its availability; what sort of benefit does its possession engender?

2. *Cost*: How costly is its employment; how expensive (complicated, difficult, resource demanding) is its use?

A natural tendency is at work in human affairs—and indeed in the dealings of rational agents generally—to keep these two items in alignment so as to maintain a proper proportioning of costs and benefits. In particular:

1. If some instrumentality affords a comparatively inexpensive means to accomplishing a needed task, we incline to make more use of it.

2. If we need to achieve a certain end often, then we try to devise less expensive ways of achieving it.

Such principles of economic rationality not only explain why people use more staples than paper clips but also account for important cognitive situations—for example, why the most frequently used words in a language tend to be among the shortest. (No ifs, ands, or buts about that!)

Economy of effort is a cardinal principle of rationality that helps to explain many aspects of the way in which we transact our cognitive business. Why are encyclopedias organized alphabetically rather than topically? Because this simplifies the search process. Why are accounts of people's doings or a nation's transactions standardly presented historically, with biographies and histories presented in chronological order? Because an account that moves from causes to effect simplifies understanding. Why do libraries group books together by topic and language rather than, say, alphabetically by author? Because this minimizes the difficulties of search and access. We are in a better position to understand innumerable features of the way in which people conduct their cognitive business once we take the economic aspect into account.

It is particularly noteworthy from such an economic point of view that there will be some conditions and circumstances in which the cost of acquiring information—even assuming that it is to be had at all in the prevailing state of the cognitive art—is simply too high relative to its value. There are (and are bound to be) circumstances in which the acquisition costs of information exceed the benefits or returns on its possession. In this regard, too, information is just like any other commodity. The price is sometimes more than we can afford and often greater than any conceivable benefit that would ensue. (This is why people generally do not count the number of hair in their eyebrows.)

Rationality and economy are thus inextricably interconnected. Rational inquiry is a matter of epistemic optimizations, of achieving the best overall balance of cognitive benefits relative to cognitive costs. Cost-benefit calculation is the crux of the economy of effort at issue. The principles of least effort—construed in a duly intellectualized manner—are bound to be a salient feature of cognitive rationality.[7] A version of Occam's Razor obtains throughout the sphere of cognitive rationality: *complicationes non multiplicandae sunt praeter necessitatem*. Efforts to secure and enlarge knowledge are worthwhile only insofar as they are cost effective in that the resources we expend for these purposes are more than compensated for through benefits obtained—as is indeed very generally the case—but not always. We are, after all, finite beings who have only limited time and energy at our disposal.

Even the development of knowledge, important though it is, is nevertheless of limited value—it is not worth the expenditure of every minute of every day at our disposal.

The standard economic process of cost-effectiveness tropism is operative throughout the cognitive domain. Rational inquiry is rigorously subject to the economic impetus to securing maximal product for minimal expenditure. Concern for answering our questions in the most straightforward, most cost-effective way is a crucial aspect of cognitive rationality in its economic dimension.

The long and short of it is that knowledge acquisition is a purposive human activity—like most of our endeavors. As such it involves the ongoing expenditure of resources for the realization of the objectives—description, explanation, prediction, and control—that represent the defining characteristics of our cognitive project. The balance of costs and benefits becomes critical here and endows the scientific enterprise with an unavoidably economic aspect.[8]

NOTES

1. *Veniet tempus quo posteri nostri tam aperta nos nescisse mirentur*: "There will come a time when our descendants will be amazed that we did not know things that are so plain to them." (Seneca, *Natural Questions,* 7.25.5.)

2. On Peirce's project on economy of research, see the author's *Peirce's Philosophy of Science* (Notre Dame and London: University of Notre Dame Press, 1976), as well as C. F. Delaney, "Peirce on 'Simplicity' and the Conditions of the Possibility of Science," in L. J. Thro, ed., *History of Philosophy in the Making* (St. Louis: University of St. Louis Press, 1974), pp. 177-94.

3. One valuable contribution in this area is Fritz Machlup, *The*

Production and Distribution of Knowledge in the United States (Princeton: Princeton University Press, 1962).

4. Fridtjof Nansen as quoted in Roland Huntford, *The Last Place on Earth* (New York: Simon & Schuster, 1985), p. 200.

5. William James, "The Sentiment of Rationality," in *The Will to Believe and Other Essays in Popular Philosophy* (New York and London: Longmans, Green & Co., 1897), pp. 78-79.

6. But is it indeed irrational to give a gift more costly than the social situation requires? By no means! It all depends on one's aims and ends, which may, on such an occasion, lie in a desire to cause the recipient surprise and pleasure, rather than merely doing the customary thing. There is an important difference between wastefulness and generosity.

7. On this theme see the classical investigations of George K. Zipf, *Human Behavior and the Principle of Least Effort* (Cambridge, Mass.: Addison Wesley, 1949). Zipf's book provides a wide variety of interesting examples of how various of our cognitive proceedings exemplify a tendency to minimize the expenditure of energy.

8. Some of this chapter's themes are also examined in chapter 8 of the author's *Cognitive Economy* (Pittsburgh: University of Pittsburgh Press, 1989).

Chapter Two

THE THEORETICAL PROSPECT
OF UNENDING SCIENCE

SYNOPSIS

*(1) Is scientific inquiry a finite project that is ulti-
mately destined to run out of new material, as
would be bound to happen with the geographic
exploration of a limited region? (2) Some theorists
respond in the negative because they see nature's
physical constitution as exhibiting an unending
complexity. (3) So drastic a stance is not needed,
however, since an unending complexity of lawful
comportment would suffice to provide for ever-new
discoveries—even with a finitely constituted domains. (4) Moreover, even a nature of simple
physical structure can engender a range of phenomena that is unendingly diverse. (5) Then too,
the basis of ongoing discovery might reside wholly
in the character of the inquiry process itself, with
responsibility for ever-present incompleteness lying totally with us inquirers rather than with
nature as such. In sum, the world's physical constitution need not be infinitely complex to sustain
the theoretical prospect of unending science.*

1. Is Scientific Discovery an
Inherently Bounded Venture?

Some years ago, the eminent biologist Bentley Glass
made newspaper headlines with his presidential

address to the American Association for the Advancement of Science, posing the question "Are there finite limits to scientific understanding, or are there endless horizons?"[1] His answer ran as follows:

> What remains to be learned may indeed dwarf imagination. Nevertheless, the universe itself is closed and finite. . . . The uniformity of nature and the general applicability of natural laws set limits to knowledge. If there are just 100, of 105, or 110 ways in which atoms may form, then when one has identified the full range of properties of these, singly and in combination, chemical knowledge will be complete. There is a finite number of species of plants and animal—even of insects—upon the earth. . . . Moreover, the universality of the genetic code, the common character of proteins in different species, the generality of cellular structure and cellular reproduction, the basic similarity of energy metabolism in all species and of photosynthesis in green plants and bacteria, and the universal evolution of living forms through mutation and natural selection all lead inescapably to a conclusion that, although diversity may be great, the laws of life, based on similarities, are finite in number and comprehensible to us in the main even now. We are like the explorers of a great continent who have penetrated to its margins in most points of the compass and have mapped the major mountain chains and rivers. There are still innumerable details to fill in, but the endless horizons no longer exist.[2]

This view of the matter sees the scientific project as an inherently bounded venture, subject to the idea that since nature is governed by a finite family of fundamental laws, it follows that in scientific inquiry, as in the geographical exploration of the planet, we are ultimately bound to reach the end of the road.[3]

There is, however, substantial warrant for thinking this idea of an inevitable end to scientific inquiry to be deeply problematic. That forecasts of the approaching completion of science are highly unrealistic becomes clear in considering some of the reasons why it would be quite wrong to expect science ever to run out of steam.

2. NATURE MIGHT EXHIBIT AN UNENDING COMPLEXITY OF PHYSICAL CONSTITUTION

To validate the prospect of endless progress in the development of natural science, some theoreticians deem it necessary to stipulate an intrinsic infinitude in the makeup of nature as a physical structure.[4] The physicist David Bohm, for example, writes: "at least as a working hypothesis science assumes the infinity of nature; and this assumption fits the facts much better than any other point of view that we know."[5] In line with this approach, Bohm postulates a principle of unending complexity in nature's makeup.

Such an infinity-of-nature postulate can take various forms. One of them considers the domain of natural laws to be inexhaustible along the lines of Pascal's idea that nature might be an endless nest of Chinese boxes, with microscopic worlds emplaced within ever more minute microcosms.[6] This approach envisions a system of wheels within wheels—somewhat on analogy with the idea that every organism comes equipped with a yet smaller organism to bite it. Sometimes this shrinkage toward the submicroscopically small is inverted into an expanding sequence moving unendingly upward toward the supratelescopically large—emplacing macroscopic worlds within ever more macroscopic ones. (Perhaps what we see as a galaxy is simply the subatomic particle of a vast macroworld, within which our whole visible universe is no more than an atom.) Following this line of thought, one might project an ascending sequence of levels—subatomic particles, atoms, molecules, organic microorganisms, animals, populations, life-systems, solar systems, galaxies, galactic clusters, worlds, world-systems, and so on—with each level furnished with its own characteristic modes of operation. Every system is potentially a member of a larger system. In either view—with ongoing increase or decrease alike—the prospect of a theoretically unending progress of science is ensured by the existence of an unending series of new worlds embracing (or embraced by) the old.[7] Science is potentially unending because there are ever-new orders of nature to be explored.

Other theorists take a less radical course and exchange the appeal to successive inclusion levels of *physical composition* for an ever-deepening succession of interaction-determinative physical forces in nature. As one physicist puts it:

> Why should nature run on just a finite number of different types of force? May there not be an infinite hierarchy of types of force just as there is an infinity of structures that may be built up of matter interacting under the influence of one or more of those forces? . . . It is perfectly possible that there exist objects that interact powerfully with each other but only exceedingly feebly with the objects with which we are familiar, that is to say with objects that interact strongly with ourselves, so that these unknown objects could build up complex structures that could share our natural world but of which we should be ignorant.[8]

Theorists of this persuasion envision an endless succession of natural forces, each operating in its own distinctive parametric sphere within an unendingly varied manifold of process.

Any such view providing for an unfolding complexity in the physical operation of nature would indeed serve to ensure that, in principle, the well of potential scientific discovery need never run dry. Such an infinite complexity of makeup, however, while sufficing to establish the cognitive inexhaustibility of nature, is by no means needed for this end—which is fortunate, since such infinity-of-nature postulates invariably have a far-fetched air about them. To see this, let us now turn our attention from the complexity of nature's *physical* makeup to the *nomic* complexity of its lawful comportment.

3. Nature Might Exhibit an Unending Complexity of Lawful Comportment

Some theorists—C. S. Peirce, for example—see the opening up of new realms of phenomena in *developmental and evolutionary* terms. Early on in world history, before the evolution of complex molecules, there was no place for biological laws, even as in the era of Neanderthal man there was no room for political economy. As the cosmos grows older, however, new modes of natural organization and process gradually evolve to afford new phenomena that are governed by emerging laws of their own. We move, say, from laws of individual physical agents to laws of variously organized complexes thereof. Since the universe affords a varied panorama of modalities of physical process evolving over time, a science that reflects this will always find new grist for its mill.

Still other theorists hold that the element of chance and randomness at work in the world of itself makes for a potentially unending hierarchical order of processes representing statistical aggregates, as it were, of lower level tendencies stabilized under the aegis of the law of large numbers. In this vein, the French physicist Jean Paul Vigier revised along Hegelian lines Pascal's earlier conception of physical nesting, reasoning as follows:

> At all levels of Nature you have a mixture of causal and statistical laws (which come from deeper or external processes). As you progress from one level to another you get new qualitative laws. Causal laws at one level can result from averages of statistical behavior at a deeper level, which in turn can be explained by deeper causal behavior, and so on *ad infinitum*. If you then admit that Nature is infinitely complex and that in consequence no final stage of knowledge can be reached, you see that at any stage of scientific knowledge causal and probability laws are necessary to describe the behavior of any phenomenon, and that any phenomenon is a combination of causal and random properties inextricably woven with one another.[9]

Such an approach envisions an unending hierarchy not of physical structures but of modes of order—of lawful comportment reflecting statistical patterns in natural phenomena.

There are two rather different sorts of complexity hierarchies. One is a hierarchy of material systems related by physical inclusion: for example, particles, atoms, molecules, macrolevel physical objects, stars and planets, galaxies, galactic clusters, etc.; or, again molecules, cells, organs, organisms, colonies, etc. Such a *physical* hierarchy of compositional order stands in contrast, however, to a law hierarchy of functional or processual order with stages moving from base-level laws that govern phenomena to higher level laws that themselves coordinate such laws, and so on. The latter sort of hierarchy of lawful order does not presuppose the former. It can go on indefinitely even in a world whose structural complexity depth is finite: complexity in point of operational functioning does not require complexity in point of physical constitution. Even a world that is finite in the structural complexity of its physical constitution may well exhibit a "hierarchy of nomic orders" in its operation, with an ongoing sequence of levels of higher order laws.

A rather fundamental principle is at work here. In a law hierarchy, any law is, potentially, a member of a wider family of laws that will itself exhibit some lawful characteristics and thus be subject to synthesis under still "higher" laws. We thus move from base-level laws governing phenomena (first-order laws) to higher level laws governing such laws (second-order laws) and so on, ascending to new levels of sophistication and comparative complexity as we move along. No matter what law may be at issue, there arise new questions about it that demand an answer in emergently new lawful terms. It becomes crucial in this context that higher level patterns are not necessarily derivable from lower level ones. The statistical frequency with which individual letters such as A and T occur in a text fails to determine the frequency with which a combination such as AT occurs. When we change the purview of our conceptual horizons, there is always in principle more to be learned—novelty that could not have been predicted from earlier, lower level information.

Consider, for example, some repeatedly exemplified physical-transaction concept and contemplate the sequence of 0's and 1's projected according to the rule that $x_i = 1$ if this applies (or is exemplified) on occasion number i, and $x_i = 0$ if not. Whenever two such concepts, C and C', generate such sequences, say

C: 0100110100 . . .

C': 1001011010 . . .

we can introduce the corresponding *matching sequence* of C and C'—namely, 0010010001 . . .—which is such that its i-th position is 1 if the two base sequences *agree* at their respective i-th positions, and 0 if they *disagree*. Such matching sequences will have a life of their own. Even if two base sequences are entirely random, their matching sequence need not be—for example, when those base sequences simply exchange 0's and 1's. (Even random phenomena can be related by laws of coordination.)

Note, moreover, that one can always regard matching sequences themselves as further base sequences, so as to yield "second-order" phenomena, as it were. One can then proceed to examine the relationships between them—or between them and other base sequences. This process yields a potential hierarchy of "laws of coordination"—at level $i + 1$ we have the laws of coordination between sequences at level i. Such a perspective illustrates how simple base phenomena can ramify to bring more and more grist to the mill of study and analysis. Increasingly sophisticated mechanisms of conceptual coordination can lead us to regard the same phenomena in the light of different complexity levels. Quite different regularities and laws can emerge at different levels.

Given such an unending exfoliation of law levels, our knowledge of the world's lawful order becomes self-potentiating, and new metadisciplines can in theory always spring up to exploit the interrelations among old disciplines. It is clear, too, that such an infinite proliferation of laws would also serve to block any prospect of completing science. There is thus no need to suppose that the *physical* complexity of nature need be unlimited for nature to have an unlimited *cognitive* depth.[10] After all, the prime task of science lies in discovering the levels of lawful order of nature, and the ongoing law complexity of nature suffices for our present purposes of providing for potentially endless discovery.

A supposition of the *structural infinitude* of nature is accordingly not needed to provide for nonterminating progress in scientific innovation. An unending depth in the *nomic complexity* of the world's code of natural law would be quite enough to underwrite the potential limitlessness of scientific discovery.

One can go still further, however, in making room for scientific innovation in a finitely complex world because—

4. NATURE MIGHT EXHIBIT AN UNENDINGLY
DIVERSITY OF PHENOMENA

Even a system that is finitely complex *both* in its physical makeup *and* in its basic law structure might nevertheless yet be infinitely complex in its *productive operations* over time. A limited producer might well engender unlimited products. Even were the number of constituents of nature to be small, the ways in which they can be combined to yield products in space-time might yet be infinitely varied. Think here of the examples of letters/syllables/words/sentences/paragraphs/ books/genres (novels, reference books, etc.) /libraries/ library systems. Even a finite nature can, like a type-writer with a limited keyboard, yield an endlessly varied text. It can produce a steady stream of new phenomena—"new" not necessarily in kind but in their functional interrelationships and thereby in their implications for theory, so that our knowledge of nature's operations can continually be enhanced and deepened. Even a relatively simple world as regards its basic operations may well exhibit an effectively infinite *cognitive* depth when one proceeds to broaden one's notion of a natural phenomenon to include not just the processes themselves and the products they produce but also the *relationships* among them.

There is, after all, no need to assume a "ceiling" to such a sequence of levels of integrative complexity of phenomenal order. Each successive level of operational or functional complexity can in principle exhibit a characteristic order of its own. The phenomena we attain at the nth level can have features whose investigation takes us to the $(n + 1)$st. New phenomena and new laws can in theory arise at every level of integrative order. The different facets of nature can generate conceptually new strata of productive operation to yield a potentially unending sequence of levels, each giving rise to its own characteristic principles of organization, themselves quite unpredictable from the standpoint of the other levels.

Suppose, for example, a natural system to be such that for essentially technical reasons a certain parameter p cannot be evaluated by us at the precise time point t but only on average during an interval around t. In such a case, the system can be very simple indeed—it need contain *no* complexities apart from those required to ensure the preceding assumptions—and yet the prospect of endless cognitive progress is nevertheless available, for as our capacity to make p determinations down to smaller and smaller time intervals increases from minutes to seconds to milliseconds to microseconds, and so on, we can obtain an increasingly comprehensive insight into the modus operandi of the system and can obtain ever fuller information about it that could not have been predetermined on the basis of earlier knowledge. Since averages at levels of larger scale do not determine those at smaller ones, quite different modes of comportment—and thus laws—could manifest themselves with the new phenomenon that arises at different levels.

Again, consider a somewhat different illustration. Let us suppose a totally random sequence of 2's and 3's, on the order of 2 3 2 2 3. . . . Suppose further a transformation that substitutes the pair 10 for 2 and 11 for 3 so as to yield:

10 11 10 10 11 . . .

We are now in a position to ascertain such "laws" as:

(1) The sequence of *even-numbered* positions 01001 . . . will be a random mix of 0's and 1's (which simply mirrors the initial random sequence of 2's and 3's)

(2) All the *odd-numbered* positions are filled by 1's.

We confront a peculiar mixture of randomness with regularity here, but of course it is only by studying *pairs* in that initial zero-one sequence that we can discern its code. Only by bringing appropriate coordination *concepts* to bear can we discern the "laws" at issue, and no matter how far we push forward along such lines, the prospect of *further* positional laws can never be eliminated—nor downplayed by claiming that such laws are inherently less significant than the rest.

There is no reason why this sort of thing cannot continue on and on, for the system always exhibits new patterns of phenomenal order over time, and so there is always more to be learned about it. There will always be new levels of functional complexity of operation to be investigated with such a system. Coordination phenomena have a life of their own. In principle, it will always be possible to discern yet further levels of structured relationship.

To be sure, it could, in theory, possibly occur that just the same relationship patterns simply recur from level to level—that the patterns of phenomena that we encounter at level $i + 1$ simply reduplicate those already met with at level i. (When it is physical configurations we are dealing with, then this situation of pattern repetition at different levels of scale will yield the "fractal" structures made prominent by E. Mandelbrot.)

This, however, is a very special case that certainly does not obtain across the board. We have no good reason to think that our world is fractal in the structure of its processes.

Even when we continue to concern ourselves with the same species of object (e.g., symbols) and the same basic laws (e.g., combination rules or grammars), we can nevertheless always in principle deepen our understanding of the phenomena by introducing increasingly powerful means of analysis to secure new sorts of data. Consider an analogy. The chess master and the beginner make exactly the same sorts of moves—the individual pieces behave exactly alike for each of them. At this level their realms "are governed by exactly the same laws." An observer exclaims: "I know all about chess, for I have now discovered the rules according to which those pieces move." Splendid! But also naive, since this business of "the basic rules of the game" is of course merely Step One. It is in point of the complexity of their governing principles that masters and beginners differ decisively.

To be sure, that chessboard is limited and finite—and so are the pieces sitting upon it. Moreover, so are the basic "laws of nature"—the moves that the different pieces may make. Nevertheless, the moves of a game of chess are covered by very different categories of "rule"—not just "the rules of the game" that define those basic legal chess moves, but rules of tactics and deeper level principles of strategy. The game can be played—and its operational phenomena thus studied—at very different levels of depth or sophistication.

As the analogy of chess makes clear, it is radically mistaken to think that we have gotten to the bottom of things when we merely grasp nature's basic physical laws. The question of what sort of game she is playing by those rules—those *basic* laws—still remains an open issue. Even though nature might be of finite physical and nomic complexity as regards its physical structure and its basic procedural laws, nevertheless it could be infinitely diverse in the unfolding operational complexity of its phenomenal products over time. To understand the world about us we need departments of biology and economics as well as departments of physics.

Regardless of the known character of a sequence of phenomena that we confront, we can never rule out the possibility that yet further patterns of relationship exist, for there will be patterns of phenomena, and patterns of such patterns, and patterns of patterns of such patterns—and on and on. We can study letter sequences as such, or move on to the level of words, and thence to sentences, and thence to paragraphs, and so to chapters, to books, to book categories, to book systems (French vs. Chinese literature), etc. Every new level of consideration will afford phenomena of its own that will themselves admit of further study and analysis. Confronted with repetitive phenomena of any description, inquiry can always in principle find new grist for its mill among the phenomena arising at higher levels of productive operation.

But is natural science not bound by a Principle of Simplicity—is it not committed to the idea that nature proceeds in fundamentally simple ways? By no means! We have no ground whatever for supposing the "simplicity" of nature. The so-called Principle of Simplicity is really a principle of complexity management:

Feel free to introduce complexity in your efforts to describe and explain nature's ways. But only when and where it is really needed. Insofar as possible "keep it simple!" Only introduce as much complexity as you really need for your scientific purposes of description, explanation, prediction, and control.

Such an approach is eminently sensible. Of course, such a principle is no more than a methodological rule of procedure for managing our cognitive affairs. Nothing entitles us to transmute this methodological precept into a descriptive/ontological claim to the effect that nature is simple—let alone of finite complexity.

* * * * *

The lesson of the preceding deliberations is clear. As best we can tell, there just is no sufficient reason within the general principles of the matter why scientific innovation need ever come to a stop. Suppose that we were to discover that:

(i) the world is of limited *descriptive complexity*: that at any given time there are only finitely many realizable states of nature

AND

(ii) the world is of limited *structural complexity*: that it is finitely stratified in the subordination and superordination of physical systems so that the world's physical constitution is inherently limited

AND

(iii) the world is of limited *nomic complexity*: that there are only finitely many ("basic") laws of nature.

Even if all this were to be so—which to all appearances it is not, and which, even were it so, we have no convincing way of establishing—still, the theoretical prospect of ongoing scientific discovery could nevertheless not be precluded. Only if we lived in a Nietzschean world of eternal recurrence—a world whose fixed patterns of occurrence reiterated themselves in a vast periodicity of unchanging repetition—could the prospect of ongoing scientific innovation be decisively abolished. There is, of course, no reason whatsoever that this prospect is even remotely plausible.

5. An Unending Prospect of Scientific Discovery Might Root Wholly in the Character of Our Inquiry Processes

There is yet another, still more fundamental aspect to the matter. It is tempting to think that the potential endlessness of scientific progress requires limitlessness on the side of its *objects*, so that the infinitude of nature must be postulated at either the structural of the nomic and/or operational level. This approach is, however, oversimplified and mistaken, for what really matters for ongoing scientific innovation is *cognitive* rather than *physical* complexity. Scientific innovation, the cognitive exploration of the ways of the world, pivots on the *interaction* of the mind with nature—on the *mind's exploitation of the data to which it gains access* for the sake of penetrating the "secrets of nature." The crucial fact is that scientific progress hinges not just on the makeup of nature herself, but also on the character of the information-acquiring processes by which we investigators investigate it.

Responsibility for the *cognitive* inexhaustibility of nature need not lie on the side of nature at all, but can in principle rest one-sidedly with us, its explorers. Even in investigating a finite system we can make infinite progress by moving ever inward from one imperfect systematization to yet another.

The introduction of *new conceptual perspectives* is the key. Each time a new discipline opens up its characteristic approach, a vantage point is gained for the reappraisal of old issues in other fields. (The body of Shakespeare's work is finite, but this does not mean that Shakespeare scholarship will come to the end of the road. Think of how psychoanalysis, for example, or sociology—disciplines that did not exist in the Bard's own day—have been used as launching platforms for innovations in Shakespearean studies.)

The complexity of exfoliating hierarchy levels need thus not inhere in the structural makeup of nature herself, but may derive from the conceptual mechanisms being deployed to study it at ever-greater depths of sophistication. The case is similar to that of the geometer, whose hierarchy of definitions, axiomatic facts, lemmas, theorems, and so on does not reside in the geometer's materials as such, but in the conceptual taxonomy that he chooses, for cognitive reasons, to impose on these materials. In and by herself, nature presumably lacks "depth" altogether, for depth (like difficulty) is an inevitably relative matter, generated through the operation of a cognitive perspective. The endless levels at issue will not be *physical* levels but *levels of consideration* that emerge from the modus operandi of inquiring beings. Complexity, after all, lies less in the objects than in the eyes of their beholder. (As John Herschel ruminated long ago, particles moving in mutual gravitational interaction are, as we human investigators see it, forever solving differential equations that, if written out in full, might circle the earth.[11])

If we are sufficiently myopic, then, even when the scene that we examine is itself only finitely complex, an ever-ampler view of it will emerge as the resolving power of our conceptual and observational instruments is increased. When we make measurements at a given level of technical sophistication, the world may appear X-wise, where X varies from level to level. At each successive state-of-the-art stage of increased precision in our investigative proceedings, the world may take on a very different nomic appearance, not because it itself *changes*, but simply because at each stage it *presents* itself differently to us.

Accordingly, the reason for the cognitive inexhaustibility of nature need not rest with nature alone. The character of our informative gathering procedures, as they come to be channeled through our conceptual/theoretical perspectives, is also bound to play no less crucial a part.[12] Innovation on the side of data can generate new theories, and new theories can transform the very meaning of the old data. Such a dialectical process of successive data/theory feedback has no inherent limits and suffices of itself to underwrite a prospect of ongoing innovation in the study of nature. We cannot write FINIS to the book of scientific inquiry as long as the prospect of a *change of mind* on our part regarding the world's workings cannot be precluded.

Considerations of this sort combine to indicate that an assumption of the structural infinity of the physical extent of the natural universe or of the functional infinity of its nomic or operational complexity is simply not required to provide for the prospect of ongoing scientific progress. Continuing discovery is quite as much a matter of how we inquirers proceed with our work as it is of the nature of the object of inquiry. The salient point is that it is *cognitive* rather than *descriptive* or *structural* or *operational* complexity that is ultimately the pivot point here and that cognitive depth need not necessarily be grounded in physical depth.

The fact of the matter is that fundamental features inherent in the structure of man's interactive inquiry into the ways of the world conspire to preclude the stabilization—let alone the definitive completion—of the processes at issue in our development of the scientific knowledge of nature. This ensures for natural science the status of a potentially unending process, for not only the procedures but also the products of scientific inquiry are inherently processual, subject to the process-characteristic phases of initiation, development, and decline. The world as science teaches us to see it as both pervasively processual and itself in process of ceaseless development.[13] This also holds for natural science itself.

Thus, the possibility of unending scientific progress is certainly there, but to assess its realizability we must turn from the theoretical to the practical level of consideration by asking: What sort of price does its cultivation extract?[14]

NOTES

1. Bentley Glass, "Science: Endless Horizons of Golden Age?", *Science*, vol. 171 (1971), pp. 23-29.

2. Ibid., p. 24

3. The geographic exploration analogy is an old standby: "Science cannot keep on going so that we are always going to discover more and more new laws. . . . It is like the discovery of America—you only discover it once. The age in which we live is the age in which we are discovering the fundamental laws of nature, and that day will never come again." (Richard Feynman, *The Character of Physical Law* [Cambridge, MA: MIT Press, 1965], 172. See also Gunter Stent, *The Coming of the Golden Age* [Garden City, NY: Doubleday, 1969]; and S. W. Hawkins, "Is the End in Sight for Theoretical Physics?," *Physics Bulletin*, vol. 32 (1981), pp. 15-17.)

4. Marxist theoreticians take this view very literally—in the manner of Lenin's idea of the "inexhaustibility" of matter in *Materialism and Empirico-Criticism* (New York: International Publishers, 1927). Purporting to inherit from Spinoza a thesis of the infinity of nature, they construe this to mean that any cosmology that denies the infinite spatial extension of the universe must be wrong.

5. "Remarks by David Bohm," in *Observation and Interpretation*, ed. by Stephan Koerner (New York and London: Macmillan, 1957), p. 56. For a fuller development of Bohm's views on the "qualitative infinity of nature," see his *Causality and Chance in Modern Physics* (London and New York: Macmillan, 1957).

6. A later writer put it this way: "Go on as far as we will, in the subdivision of continuous quantity, yet we never get down to the absolute point. Thus scientific method leads us to the inevitable conception of an infinite series of successive orders of infinitely small quantities. If so, there is nothing impossible in the existence of a myriad universe within the compass of a needle's point, each with its stellar systems, and its suns and planets, in number and variety unlimited." (W. S. Jevons, *The Principles of Science*, 2nd ed. [London: Macmillan & Co., 1877], p. 767.)

7. The idea of a succession of "layers of depth" in the analysis of nature, each giving rise to its own characteristic body of laws in such a way that each of these law-manifolds is sequentially deeper penetration of the structure of nature was, as far as I know, first mooted in contemporary physics by E. P. Wigner, "The Limits of Science,"

Proceedings of the American Philosophical Society, vol. 94, (1950), 422-27.   It had already been contemplated by Charles Sanders Peirce, however, half a century earlier.

8. Sir Denys Wilkinson, *The Quarks and Captain Ahab* or: *The Universe as Artifact*, (Stanford: Stanford University Press, 1977; Schiff Memorial Lecture), pp. 4-5.

9. "The Concept of Probability in the Frame of the Probabilistic and the Causal Interpretation of Quantum Mechanics," in Stephan Koerner (ed.), *op. cit.*, p. 77.   Cf. also E. P. Wigner in "The Limits of Science," *op cit.*

10. For an interesting and suggestive analysis of "the architecture of complexity," see Herbert A. Simon, *The Sciences of the Artificial* (Cambridge, MA: MIT Press, 1969).

11. John Herschel, *Familiar Lectures on Scientific Subjects* (London: A. Strahan, 1867), p. 458.

12. This idea that our knowledge about the world reflects an *interactive* process, to which both the object of knowledge (the world) and knowing subject (the inquiring mind) make essential and ultimately inseparable contributions is elaborated in the author's books *Conceptual Idealism* (Oxford: Basil Blackwell, 1973) and *Methodological Pragmatism* (Oxford: Basil Blackwell, 1977).

13. Compare D. A. Bromley's observation: "Even if physicists could be sure that they had identified all the particles that can exist, some obviously fundamental question would remain.   Why, for instance, does a certain universal ratio in atomic physics have the particular value 137.036 and not some other value?   This is an experimental result: the precision of the experiments extends today to these six figures. Among other things, this number relates the extent or size of the electron to the size of the atom, and that in turn to the wavelength of light emitted.   From astronomical observation it is known that this fundamental ratio has the same numerical value for atoms a billion years away in space and time.   As yet there is no reason to doubt that other fundamental ratios, such as the ratio of the mass of the proton to that of the electron, are as uniform throughout the universe as is the geometrical ratio pi equals 3.14159. Could it be that such physical ratios are really, like pi, mathematical aspects of some underlying logical structure? If so, physicists are not much better off than people who must resort to wrapping a string around a cylinder to determine the value of pi!   For theoretical physics thus far shed hardly a glimmer of light on this question." (D.

A. Bromley, et al., *Physics in Perspective*, Student Edition (Washington, D.C. National Academy of Science/National Research Council Publications, 1973), p. 28.

14. This chapter's themes are also addressed in Chapter 4 ("An End to Science?") of the author's collection, *Forbidden Knowledge* (Dordrecht, Holland: D. Reidel, 1987).

Chapter Three

TECHNOLOGICAL ESCALATION AND THE EXPLORATION MODEL OF NATURAL SCIENCE

SYNOPSIS

(1) The development of inquiry in natural science is best understood on the analogy of exploration— to be sure, not in the geographical mode but rather exploration in nature's parametric space of such physical quantities as temperature, pressure, and field strength. (2) The technology-mediated exploration that comes into play here involves inter- actions between man and nature that become increasingly difficult (and expensive) as we move ever farther away from the home base of the accus- tomed environment of our evolutionary heritage. (3) The course of scientific progress accordingly involves a technological escalation—an ascent to successively higher levels of technological sophis- tication that is unavoidably required for the pro- duction of duly informative observational data.

1. THE EXPLORATION MODEL OF SCIENTIFIC INQUIRY

In theory, a prospect of unending scientific progress lies before us, but its practical realization is some- thing else again. One of the most striking and important facts about scientific research is that the ongoing resolution of significant new questions faces increasingly high demands for the generation and cognitive exploitation of data. Though the veins of cognitive gold run on, they become increasingly difficult—and expensive—to mine.

In developing natural science, we humans began by exploring the world in our own locality, and not just our spatial neighborhood but—more far-reachingly—our *parametric* neighborhood in the space of physical variables such as temperature, pressure, and electric charge. Near the "home base" of the state of things in our accustomed natural environment, we can operate with relative ease and freedom—thanks to the evolutionary attunement of our sensory and cognitive apparatus—in scanning nature with the unassisted senses for data regarding its modes of operation. In due course, however, we accomplish everything that can be managed by these straightforward means. To do more, we have to extend our probes into nature more deeply, deploying increasing technical sophistication to achieve more and more demanding levels of interactive capability. We have to move ever further away from our evolutionary home base in nature toward increasingly remote observational frontiers. From the egocentric standpoint of our local region of parameter space, we journey ever more distantly outward to explore nature's various parametric dimensions in the search for cognitively significant phenomena.

The appropriate picture is not, of course, one of geographical exploration but rather of the physical exploration—and subsequent theoretical systematization—of phenomena distributed over the parametric space of the physical quantities spreading out all about us.

This approach in terms of exploration provides a conception of scientific research as a prospecting search for the new phenomena demanded by significant new scientific findings. As the range of telescopes, the energy of particle accelerators, the effectiveness of low-temperature instrumentation, the potency of pressurization equipment, the power of vacuum-creating contrivances, and the accuracy of measurement apparatus increases—that is, as our capacity to move about in the parametric space of the physical world is enhanced—new phenomena come into view. After the major findings accessible via the data of a given level of technological sophistication have been achieved, further major findings become realizable only when one ascends to the next level of sophistication in data-relevant technology. Thus the key to the great progress of contemporary physics lies in the enormous strides that an ever more sophisticated scientific technology has made possible through enlarging the observational and experimental basis of our theoretical knowledge of natural processes.[1]

In cultivating scientific inquiry, we scan nature for interesting phenomena and grope about for the explanatorily useful regularities they may suggest. As a fundamentally inductive process, scientific theorizing calls for devising the least complex theory structure capable of accommodating the available data. At each stage we try to embed the phenomena and their regularities within the simplest (cognitively most efficient) explanatory structure able to answer our questions about the world and to guide our interactions in it. Step by step as the process advances, however, we are driven to further, ever-greater demands that can be met only with a yet more powerful technology of data exploration and management.

This idea of the exploration of parametric space provides a basic model for understanding the mechanism of scientific innovation in mature natural science. New technology increases our range of access within the parametric space of physical processes. Such increased access brings new phenomena to light, and the examination and theoretical accommodation of these phenomena are the basis for growth in our scientific understanding of nature.

2. THE DEMAND FOR ENHANCEMENT

Natural science is fundamentally empirical, and its advance is critically dependent not on human ingenuity alone but on the monitoring observations to which we can gain access only through interactions with nature. The days are long past when useful scientific data can be had by unaided sensory observation of the ordinary course of nature. Artifice has become an indispensable route to the acquisition and processing of scientifically useful data. The sorts of data on which scientific discovery nowadays depends can be generated only by technological means.

The pursuit of natural science as we know it embarks us on a literally endless endeavor to improve the range of effective experimental intervention, because only by operating under new and heretofore inaccessible conditions of observational or experimental systemization —attaining extreme temperature, pressure, particle velocity, field strength, and so on—can we realize situations that enable us to put knowledge-expanding hypotheses and theories to the test. The enormous power, sensitivity, and complexity deployed in present-day experimental science have not been sought for their own sake but rather because the research frontier has moved on into an area where this sophistication is the indispensable requisite of further progress. In science, as in war, the battles of the present cannot be fought effectively with the armament of the past. As one acute observer has rightly remarked: "Most critical experiments [in physics] planned today, if they had to be constrained within the technology of even ten years ago, would be seriously compromised."[2]

With the enhancement of scientific technology, the size and complexity of this body of data inevitably grow, expanding in quantity and diversifying in kind. Technological progress constantly enlarges the window through which we look out upon nature's parametric space. In developing natural science, we continually enlarge our view of this space and then generalize upon what we see. What we have here, however, is not a homogeneous lunar landscape, where once we have seen one sector we have seen it all and where theory projections from lesser data generally remain in place when further data comes our way. Historical experience shows that there is every reason to expect that our ideas about nature are subject to constant radical changes as we explore parametric space more extensively. The technologically mediated entry into new regions of parameter space constantly destabilizes the attained equilibrium between data and theory. It is the engine that moves the frontier of progress in natural science.

3. TECHNOLOGICAL ESCALATION:
AN ARMS RACE AGAINST NATURE

This situation points toward the idea of a "technological level," corresponding to a certain state of the art in the technology of inquiry in regard to data generation and processing. Our technology of inquiry falls into relatively distinct levels or stages in sophistication—correlatively with successively "later generations" of instrumentation and manipulative machinery. These levels are generally separated from one another by substantial (roughly, order-of-magnitude) improvements in performance in regard to such information-providing parameters as measurement exactness, data-processing volume, detection sensitivity, high voltages, high or low temperatures, and so on. The perspective afforded by such a model of technologically mediated prospecting indicates that progress in natural science has heretofore been relatively easy because in this earlier explanation of our parametric neighborhood we have been able—thanks to the evolutionary heritage of our sensory and conceptual apparatus—to operate with relative ease and freedom in exploring our own parametric neighborhood in the space of physical variables such as temperature, pressure, radiation, and so on. Scientific innovation, however, becomes even more difficult—and expensive as we push out further from our home base toward the more remote frontiers.

No doubt, nature is in itself uniform as regards the distribution of its diverse processes across the reaches of parameter space. It does not favor us by clustering them in our accustomed parametric vicinity: significant phenomena do not dry up outside our parochial neighborhood, but scientific innovation becomes more and more difficult—and expensive—as we push our explorations even further away from our evolutionary home base toward increasingly remote frontiers.

Too, phenomenological novelty is seemingly inexhaustible: we can never feel confident that we have gotten to the bottom of it. Nature always has fresh reserves of phenomena at her disposal, hidden away in those ever more remote regions of paramative space. Successive stages in the technological state of the art of scientific inquiry accordingly lead us to ever-different views about the nature of things and the character of their laws. Without an ever-developing technology, scientific progress would soon grind to a halt. The discoveries of today cannot be made with yesterday's equipment and techniques. To conduct new experiments, to secure new observations, and to detect new phenomena, an ever more powerful investigative technology is needed. Scientific progress depends crucially and unavoidably on our technical capability to penetrate into the increasing distant—and increasingly difficult—reaches of the spectrum of physical parameters in order to explore and to explain the ever more remote phenomena encountered there.

The salient characteristic of this situation is that, once the major findings accessible at a given level of sophistication in data technology have been attained, further major progress in any given problem area requires ascent to a higher level on the technological scale. Every data-technology level is subject to discovery saturation, but the exhaustion of prospects at a given level does not, of course, bring progress to a stop.

Once the potential of a given state-of-the-art level has been exploited, not all our piety or wit can lure the technological frontier back to yield further significant returns at this stage. Further substantive findings become realizable only by ascending to the next level of sophistication in data-relevant technology. The exhaustion of the prospects for data extraction at a given data-technology level does not, of course, bring progress to a stop. Rather, the need for enhanced data forces one to look further and further from man's familiar "home base" in the parametric space of nature. Thus, while scientific progress is in principle always possible—there being no absolute or intrinsic limits to significant scientific discovery—the *realization* of this ongoing prospect demands a continual improvement in the technological state of the art of data extraction or exploitation.

The requirement for technological progress to advance scientific knowledge has far-reaching implications for the nature of the enterprise.

The instrumentalities of scientific inquiry can be enhanced not only on the side of theoretical resources but preeminently on the side of the technological instrumentalities of observational and experimental intervention. Pioneering scientific research will always operate at the technological frontier, and nature inexorably exacts a drastically increasing effort with respect to the acquisition and processing of data for revealing her "secrets." This accounts for the recourse to more and more sophisticated technology for research in natural science.

The increasing technological demands that are requisite for scientific progress, however, mean that each step ahead gets more complex and more expensive as those new parametric regions grow increasingly remote. With the progress of science, nature becomes less and less yielding to the efforts of further inquiry. We are faced with the need to push nature harder and harder to achieve cognitively profitable interactions. The dialectic theory and experiment carry natural science ever deeper into the range of greater costs.

We thus arrive at the phenomenon of *technological escalation*. The need for new data forces us to look further and further from man's familiar home base in the parametric space of nature. Thus, while scientific progress is in principle always possible—there being no absolute or intrinsic limits to significant scientific discovery—the realization of this ongoing prospect demands a continual enhancement in the technological state of the art of data extraction or exploitation.

The perspective afforded by such a process of escalation indicates that progress in natural science was at first relatively undemanding because we have explored nature in our own parametric neighborhood.[3] In addition, seeing that the requisite technology was relatively crude, economic demands were minimal at this stage. Over time, however, an ongoing escalation in the resource costs of significant scientific discovery arose from the increasing technical difficulties of realizing this objective, difficulties that are a fundamental —and an ineliminable—part of an enterprise of empirical research, for we must here contrive ever more "far-out" interactions with nature, operating in a continually more difficult, accordingly, sector of parametric space.

Nature becomes less and less yielding to the efforts of our inquiry. As science advances, we are faced with the need to push nature harder and harder to achieve cognitively profitable interactions. That there is "pay dirt" deeper down in the mine avails us only if we can actually dig there. New forces, for example, may well be in the offing, if one able physicist is right:

> We are familiar, to varying degrees, with four types of force: gravity, electricity, the strong nuclear force that holds the atomic nucleus together and the weak force that brings about radioactive decay by the emission of electrons. . . . Yet it would indeed be astonishing if . . . other types of force did not exist. Such other forces could escape our notice because they were too weak to have much distinguishable effect or because they were of such short range that, no matter whether they were weak or not, the effects specifically associated with their range were contained within the objects of the finest scale that our instruments had so far permitted us to probe.[4]

Of course, such weak forces would enter into our picture of nature only if our instrumentation was able to detect them. This need for the constant enhancement of scientifically relevant technology lies at the basis of the enormous increase in the human and material resources needed for modern experimental science.

Given that we can learn about nature only by interacting with it, Newton's third law of countervailing action and reaction becomes a fundamental principle of epistemology. Everything depends on just how and *how hard* we can push against nature in situations of observational and detectional interaction. As Bacon saw, nature will never tell us more than we can forcibly extract from her with the means of interaction at our disposal. Also, what we can manage to extract by successively deeper probes is bound to wear a steadily changing aspect, because we operate in new circumstances where old conditions cannot be expected to prevail and the old rules no longer apply.

The idea of scientific progress as the correlate of a movement through sequential stages of technological sophistication was already clearly discerned by the astute Charles Sanders Peirce around the turn of the century:

> Lamarckian evolution might, for example, take the form of perpetually modifying our opinion in the effort gradually to make that opinion represent the known facts as more and more observations come to be collected. . . . But this is not the way in which science mainly progresses. It advances by leaps; and the impulse for each leap is either some new observational resource, or some novel way of reasoning about the observations. Such a novel way of reasoning might, perhaps, also be considered as a new observational means, since it draws attention to relations between facts which would previously have been passed by unperceived.[5]

This circumstance has far-reaching implications for the perfectability of science. The impetus to augment our science demands an unremitting and unending effort to enlarge the domain of effective experimental intervention, for only by operating under new and heretofore inaccessible conditions of observational or experimental systematization—attaining ever more extreme temperature, pressure, particle velocity, field strength, and so on—can we bring new grist to our scientific mill.

To be sure, there is, on its basis, no inherent limit to the possibility of future progress in scientific knowledge, but the exploitation of this theoretical prospect gets ever more difficult, expensive, and demanding in terms of effort and ingenuity. With scientific progress we are engaged in a situation of price inflation where further equal-size steps become more difficult with every step we take. New findings of equal level of significance require ever-greater aggregate efforts. In the ongoing course of scientific progress, the earlier investigations in the various departments of inquiry are able to skim the cream, so to speak: they take the "easy pickings," and later achievements of comparable significance require ever-deeper forays into complexity and call for ever-increasing bodies of information. (It is important to realize that this cost increase is not because latter-day workers are doing *better* science, but simply because it is harder to achieve *the same level* of science: one must dig more deeply or search more widely to achieve results of the same significance as before.)

Physicists often remark that the development of our understanding of nature moves through successive layers of *theoretical* sophistication.[6] Scientific progress, however, is clearly no less dependent on continual improvements in strictly *technical* sophistication:

Some of the most startling technological advances in our time are closely associated with basic research. As compared with 25 years ago, the highest vacuum readily achievable has improved more than a thousand-fold; materials can be manufactured that are 100 times purer; the submicroscopic world can be seen at 10 times higher magnification; the detection of trace impurities is hundreds of times more sensitive; the identification of molecular species (as in various forms of chromatography) is immeasurably advanced. These examples are only a small sample. . . . Fundamental research in physics is crucially dependent on advanced technology, and is becoming more so.[7]

Without an ever-developing technology, scientific progress would grind to a halt. The discoveries of today cannot be advanced with yesterday's instrumentation and techniques. To secure new observations, to test new hypotheses, and to detect new phenomena, an ever more powerful technology of inquiry is needed. Throughout the natural sciences, technological progress is a crucial requisite for cognitive progress.

Frontier research is true *pioneering*: what counts is not just doing it but doing it *for the first time*. Aside from the initial reproduction of claimed results needed to establish the reproducibility of results, repetition in *research* is in general pointless. As one acute observer has remarked, one can follow the diffusion of scientific technology "from the research desk down to the school-room":

The emanation electroscope was a device invented at the turn of the century to measure the rate at which a gas such as thorium loses its radioactivity. For a number of years it seems to have been used only in the research laboratory. It came into use in instructing graduate students in the mid-1930's, and in college courses by 1949. For the last few years a cheap commer-

cial model has existed and is beginning to be introduced into high school courses. In a sense, this is a
victory for good practice; but it also summarizes the sad
state of scientific education to note that in the research
laboratory itself the emanation electroscope has long
since been removed from the desk to the attic.[8]

In science, as in a technological arms race, one is
simply never called on to keep doing what was done
before. And ever more challenging task is posed by the
constantly *escalating* demands of science for the enhanced data that can be obtained only at the increasingly costly new levels of technological sophistication.
One is always forced further up the mountain, ascending to higher levels of technological performance—and
of expense. As science endeavors to extend its "mastery over nature," it thereby comes to be involved in a
technology-intensive arms race against nature, with
all of the practical and economic implications characteristic of such a process.

The enormous power, sensitivity, and complexity deployed in present-day experimental science have not
been sought for their own sake but rather because the
research frontier has moved on into an area where
added sophistication is the indispensable requisite of
ongoing progress. In science, as in war, the battles of
the present cannot be fought effectively with the ar-
maments of the past.[9]

NOTES

1. A homely fishing analogy of A. S. Eddington's is useful here. He
saw the experimentalists as akin to a fisherman who trawls nature
with the net of his equipment for detection and observation. Now
suppose (says Eddington) that a fisherman trawls the seas using a

fishnet of two-inch mesh. Then fish of a smaller size will simply go uncaught, and those who analyze the catch will have an incomplete and distorted view of aquatic life. The situation in science is the same. Only by improving our observational means of trawling nature can such imperfections be mitigated. (See A. S. Eddington, *The Nature of the Physical World* [New York: A M S Press, 1928]).

2. D. A. Bromley et al., *Physics in Perspective, Student Edition* (Washington, D.C.: National Academy of Sciences, 1973), pp. 16, 13. See also Gerald Holton, "Models for Understanding the Growth and Excellence of Scientific Research," in Stephen R. Graubard and Gerald Holton, eds., *Excellence and Leadership in a Democracy* (New York: Columbia University Press, 1962), p. 115.

3. Note, however, that an assumption of the finite dimensionality of the phase space of research-relevant physical parameters becomes crucial here, for if these were limitless in number, one could always move on to the inexpensive exploitation of virgin territory.

4. Sir Denys H. Wilkinson, *The Quarks and Captain Ahab or: The Universe as Artifact* (Stanford: Stanford University Press, 1977; Schiff Memorial Lecture), pp. 12-13.

5. *Collected Papers*, ed. by C. Hartshorne et al., Vol. I (Cambridge, Mass.: Harvard University Press, 1931), pp. 44-45 (sects. 108-109).

6. "Looking back, one has the impression that the historical development of the physical description of the world consists of a succession of layers of knowledge of increasing generality and greater depth. Each layer has a well defined field of validity; one has to pass beyond the limits of each to get to the next one, which will be characterized by more general and more encompassing laws and by discoveries constituting a deeper penetration into the structure of the Universe than the layers recognized before." (Edoardo Amaldi, "The Unity of Physics," *Physics Today*, vol. 261, no. 9 [September 1973], p. 24.) See also E. P. Wigner, "The Unreasonable Effectiveness of Mathematics in the Natural Sciences, "*Communication on Pure and Applied Mathematics*, vol. 13 (1960), pp. 1-14; as well as his "The Limits of Science," *Proceedings of the American Philosophical Society*, vol. 93 (1949), 521-526. Compare also Chapter 8 of Henry Margenau, *The Nature of Physical Reality* (New York: McGraw Hill, 1950).

7. D. A. Bromley et al., *Physics in Perspective,* p. 23.

8. Gerald Holton, "Models for Understanding the Growth and Excellence of Scientific Research," in Stephen R. Graubard and Gerald Holton, eds., *Excellence and Leadership in a Democracy* (New York: Columbia University Press, 1962), p. 115.

9. Some of the themes of this chapter are also addressed in Chapter 7, "Cost Escalation in Empirical Inquiry" of the author's *Cognitive Economy* (Pittsburgh: University of Pittsburgh Press, 1989). His *Scientific Progress* (Oxford: Basil Blackwell, 1978) and *The Limits of Science* (Berkeley and Los Angeles: University of California Press, 1984) are also relevant.

Chapter Four

SCIENTIFIC PROGRESS
IS NOT ASYMPTOTIC

SYNOPSIS

(1) Scientific inquiry seeks to develop a harmoni-ously systematized coordination of theorizing con-jecture with the determinable data. However, this attempt at equilibriation repeatedly sustains the destabilizing shocks of enlarged experience. Our technologically mediated entry into new regions of parameter space confronts us with this task in ever-renewed forms. (2) In theory, the prospect of such ongoing "scientific revolutions" is potentially unending, and there is no warrant for a theory of convergence that sees the innovations of theorizing science as being of constantly diminishing scope; with discoveries in natural science, later does not mean lesser. The increasing cost of further pro-gress is, however, a portentous consideration.

1. THEORIZING AS INDUCTIVE PROJECTION

Even though neither present nor future science manages to depict reality with definitive finality, perhaps there is nevertheless a gradual *convergence* toward a definitively true account of nature at the level of scientific theorizing.

Considerations of great principles indicate that this is very implausible. Our exploration of physical parameter space is inevitably incomplete. We can never exhaust the whole range of temperatures, pressures, particle velocities, etc., and so we inevitably face the (very real) prospect that the regularity structure of the as yet inaccessible cases will not conform to the (generally simpler) patterns of regularity prevailing in the presently accessible cases. By and large, future data do not accommodate themselves to present theories. Newtonian calculations worked marvelously for predicting solar-system phenomenology (eclipses, planetary conjunctions, and the rest), but this does not show that classical physics has no need for fundamental revision.

Scientific theory formation is, in general, a matter of spotting a local regularity of phenomena in parametric space and then projecting it "across the board," maintaining it globally. The theoretical claims of science are themselves never local—they are not spatiotemporally local and they are not parametrically local either. They stipulate—quite ambitiously—how things are always and everywhere. It does not require a sophisticated knowledge of the history of science to realize that our worst fears are usually realized—that it is seldom if ever the case that our theories survive intact in the wake of substantial extensions in our access to sectors of parametric space. The history of science is a history of episodes of leaping to the wrong conclusions.

Theorizing in natural science is a matter of triangulation from observations—of inductive generalization from the data. Sensibly enough, induction as a rational process of inquiry constructs the simplest, most economical cognitive structures to house these data comfortably. It calls for searching out the simplest pattern of regularity that can adequately accommodate our data regarding the issues at hand and then projecting them across the entire spectrum of possibilities in order to answer our general questions. Accordingly, scientific theorizing, as a fundamentally inductive process, involves the search for, or the construction of, the least complex theory structure capable of accommodating the available body of data —proceeding under the aegis of established principles of inductive systematization: uniformity, simplicity, harmony, and such principles that implement the general idea of cognitive economy. Directly evidential considerations apart, the warrant of inductively authorized contentions turns exactly on this issue of the efficient and effective accomodation of the data— on consilience, mutual interconnection, and systemic enmeshment. Induction is a matter of building up the simplest structure capable of "doing the job." The key principle is that of simplicity and the ruling injunction that of cognitive economy. Complications cannot be ruled out, but they must always pay their way in terms of increased systemic adequacy.

Simplicity and generality are the cornerstones of inductive systematization. One very important point must be stressed in this connection. Scientific induction's basic idea of a *coordinative systematization of question-resolving conjecture with the data of experience* may sound like a very conservative process, but this impression would be quite incorrect. The drive to systematization embodies an imperative to broaden the range of our experience—to extend and to expand insofar as possible the data base from which our theoretical triangulations proceed. In the design of cognitive systems, implicity/harmony and comprehensive-ness/inclusiveness are two components of one whole. The impetus to ever-ampler comprehensiveness indi-cates why the ever-widening exploration of nature's parameter space is an indispensable part of the process.

Progress in natural science is a matter of dialogue or debate between theoreticians and experimentalists. The experimentalists probe nature to see its reactions, to seek out phenomena, and the theoreticians take the resultant data and weave a theoretical fabric about them. Seeking to devise a framework of rational un-derstanding, they construct their explanatory models to accommodate the findings that the experimentalists put at their disposal. But once the theoreticians have had their say, the ball returns to the experimentalists' court. Employing new, more powerful means for prob-ing nature, they bring new phenomena to view, new data for accommodation. Precisely because these data are new and inherently unpredictable, they often fail to fit the old theories. Theory extrapolations from the old data could not encompass them; the old theories do not accommodate them. Thus, a disequilibrium arises between existing theory and new data, and at this stage, the ball reenters the theoreticians' court. New theories must be devised to accommodate the new, nonconforming data, so the theoreticians set about weaving a new theoretical structure to accommodate the new data. They endeavor to restore the equilib-rium between theory and data once more. When they succeed, the ball returns to the experimentalists' court, and the whole process starts over again.

In pursuing the venture of scientific inquiry, we scan nature for interesting phenomena and project explanatorily useful regularities to try to reach further. Here breadth of coverage in point of data and economy of means in point of theory are our guiding stars, but of course things can go wrong. As regards both the observable regularities of nature and the discernible constituents of nature, very different results emerge at various levels of capacity in the observational state of the art. An analogy may prove helpful. Suppose we initially investigate a type of object X. Proceeding at the first level of sophistication, we see it as constituted of minute components whose structure is spherical. On closer investigation, however, we find (at the next level of sophistication) that these "component parts" were not actually units, but mere constellations, mere clouds of small specks. When we investigate still more deeply, it emerges that the component specks that constitute these "clouds" themselves have a rectangular form. Suppose further that at the next level those rectangular configurated "components" themselves emerge as mere constellations, composed of even more triangular constituents, and so on.

As this analogy indicates, physical nature can exhibit a very different aspect when viewed from the vantage point of different levels of sophistication in the technology of nature-investigator interaction, and this possibility is in fact realized. As ample experience indicates, every new stage of investigative sophistication brings to the fore a different order or aspect of things with respect to the modus operandi of nature. What we find in investigating nature always in some degree reflects the character of our technology of observation; what we can detect, or find, in nature is always something that depends on the mechanisms by which we search. And more sophisticated searches invariably engender changes of mind.

2. LATER NEED NOT BE LESSER

An insidiously alluring argument arises at this point. It contends that limited access to new phenomena does not matter for scientific progress because these further data would in any case yield only minor corrections located more decimal places out. The view that underlies such a position is that further changes are smaller changes: that a juncture has been reached where the additional advances of science do no more than provide further minor details and readjustments in a basically completed picture of how nature functions: that what we don't yet know doesn't matter all that much.

One particularly interesting and well-developed statement of this point of view is that of the biologist Gunther S. Stent, *The Coming of the Golden Age: A View of the End of Progress*:

> I want to consider what I believe to be intrinsic limits to the science, limits to the accumulation of meaningful statements about the events of the outer world. I think everyone will readily agree that there are *some* scientific disciplines which, by reason of the phenomena to which they purport to address themselves, are *bounded*. Geography, for instance, is bounded because its goal of describing the features of the Earth is clearly limited. . . . And, as I hope to have shown in the preceding chapters, genetics is not only bounded, but its goal of understanding the mechanism of transmission of hereditary information *has*, in fact, been all but reached. . . . [To be sure] the domain of investigation

of a bounded scientific discipline may well present a vast and practically inexhaustible number of events for study. But the discipline is bounded all the same because its goal is in view. . . . There is at least one scientific discipline, however, which appears to be *openended*, namely physics, or the science of matter. . . . But even to encounter limitations in practice . . . [for] there are purely physical limits to physics because of man's own boundaries of time and energy. These limits render forever impossible research projects that involve observing events in regions of the universe more than ten or fifteen billion light-years distant, traveling very far beyond the domain of our solar system, or generating particles with kinetic energies approaching those of highly energetic cosmic rays.[1]

In such a view, natural science is approaching the end of its tether.

Along these lines we also have the precedent of Charles Sanders Peirce's idea of *convergent approximation*.[2] This calls for envisaging a situation where, with the passage of time, the results we reach grow increasingly concordant. In the face of such a course of successive changes of ever-diminishing significance, we could proceed to maintain that the world may not really be what *present* science claims it to be, but rather is as the ever more clearly emerging science-in-the-limit claims it to be. The reality of ongoing changes is now unimportant because with the passage of time those changes matter less and less. We increasingly approximate an essentially stable picture.

This prospect is certainly a theoretically possible one, but neither historical experience nor considerations of general principle provide reason to think that it is a real possibility. Rather, one does well here to adopt the view that Stanley Jevons articulated more than a century ago:

> In the writings of some recent philosophers, especially of Auguste Comte, and in some degree John Stuart Mill, there is erroneous and hurtful tendency to represent our knowledge as assuming an approximately complete character. At least these and many other writers fail to impress upon their readers a truth which cannot be too constantly borne in mind, namely, that the utmost successes which our scientific method can accomplish will not enable us to comprehend more than an infinitesimal fraction of what doubtless is to comprehend.[3]

Nothing has happened in the interim to lead one to dissent from these strictures. We cannot realistically expect that our science, at *any* given stage of its actual development, will ever be in a position to afford us more than a very partial and incomplete access to the phenomena of nature, to reemphasize: in natural science, imperfect *physical* control is bound to mean imperfect *cognitive* control.

To evaluate this prospect, it is useful to return to the previously described exploration model of scientific progress as a matter of exploring our *parametric* neighborhood of physical variables such as temperature, pressure, field strength, and so on. We must acknowledge here that it would be bizarre indeed if significant phenomena were to dry up as we move beyond our immediate parametric neighborhood. Clearly, nature distributes its various processes across *all* the reaches of parameter space and does not favor us by clustering them in our parametric vicinity. Indeed, all historical experience counterindicates this.

Any theory of convergence in science, however carefully crafted, will shatter under the impact of the *conceptual innovation* that becomes necessary to deal with the new phenomena encountered the wake of technical escalation. Such innovation continually brings entirely new, radically different scientific concepts to the fore, carrying in its wake an ongoing wholesale revision of "established fact." Investigators of the physical phenomena of an earlier era not only did not *know* what the half-life of californium was, but they would not have *understood* it even if this fact had been explained to them. This aspect of the matter deserves closer attention.

"Which of the four elements (air, earth, fire, water) is the paramount *'archê'*, the fundamental type of stuff from which the whole of physical reality originates?" asked the early Milesians in pre-Socratic Greece. They contemplated just those four alternatives—together with the fifth possibility of a neutral, intermediate stuff. It did not occur to them that their whole inquiry was abortive because it was based on a misguided conception of "elements," nor did it appear to be a realistic prospect to all those late nineteenth-century physicists who investigated the properties of the luminiferous aether that no such medium for the transmission of light and electromagnetism might exist at all.

In factual inquiry into the ways of the world we can do no better than to pose questions and canvass the currently visible alternatives, but the questions we can pose are limited by our conceptual horizons, and the answers we can envision are also limited by the cognitive state of the art. (The Greeks could not have asked about continental drift; the Romans could not have thought of explaining the tides through gravitation.) Of course, the whole process of canvassing answers can come to grief because the very question being asked is based on untenable suppositions.

Ongoing scientific progress is not simply a matter of increasing accuracy by extending the numbers at issue in our otherwise stable descriptions of nature out to a few more decimal places. Significant scientific progress is genuinely revolutionary in involving a *fundamental change of mind* about how things happen in the world. Progress of this caliber is generally a matter not of adding further facts—on the order of filling in a crossword puzzle. It is, rather, a matter of changing the very framework itself. And this fact blocks the theory of convergence.

Even extraordinary accuracy with respect to the entire range of *currently manageable* cases does not betoken actual correctness—it merely reflects adequacy over that limited range. No matter how far we broaden that "limited range of 'presently accessible' cases," we still achieve no assurance (or even probability) that a theory corpus that accommodates (perfectly well) the range of "presently achievable outcomes" will hold across the board. The upshot is that both as regards the observable *regularities* of nature and the discernable *constituents* of nature, very different results that project very different views of the situation can—and almost invariably do—emerge at successive levels of the observational state of the art. Almost invariably we deal at every stage with a different order or aspect of things. The reason why nature exhibits different aspects at different levels is not that nature herself is somehow stratified and has different levels of being or of operation but rather that (1) the character of the available nature-investigative interactions is variable and differs from level to level, and (2) the character of the "findings" at which one arrives will hinge on the character of these nature-investigative interactions. For—to reemphasize—what we detect or "find" in nature is always something that depends on the mechanisms by which we search. The phenomena we detect will depend not merely on nature's operations alone, but on the physical and conceptional instruments we use in probing them.

In any convergent process, *later* is *lesser*, but since scientific progress on matters of fundamental importance is generally a matter of replacement rather than mere supplementation, there is no good reason for seeing the *later* findings of science as lesser than the significance of their bearing within the cognitive enterprise—to think that nature will be cooperative in always yielding its most important secrets early on and reserving nothing but the relatively insignificant for later on (nor does it seem plausible to think of nature as perverse, luring us ever more deeply into deception as inquiry proceeds.) A very small-scale effect at the level of phenomena—even one that lies very far out along the extremes of a "range exploration" in terms of temperature, pressure, velocity, or the like—can force a far-reaching revolution and have a profound impact by way of major theoretical revisions. (Think of special relativity in relation to aether-drift experimentation, or general relativity in relation to the perihelion of Mercury.)

Given that natural science progresses mainly by substitutions and replacements that involve comprehensive overall revisions of our picture of the processes at issue, it seems sensible to say that the shifts across successive scientific "revolutions" maintain the same level of overall significance when taken as a whole. At the cognitive level, a scientific innovation is simply a matter of change. Scientific progress is neither a convergent nor a divergent process.[4] Successive stages in the technological state of the art of scientific inquiry lead us to different views about the nature of things and the character of their laws, and at each level of technical sophistication we get a substantially different overall story—with differences that by no means shrink to a vanishing point as the process moves along. Later is not lesser; it is just different. Theoretical progress in science is essentially a matter of substantiation. (As in politics, one cannot replace something with nothing.)

On this basis, one arrives at a view of scientific progress as confronting us with a situation in which every major successive stage in the evolution of science yields innovations, and these innovations are—on the whole—of roughly equal overall interest and importance. Accordingly, there is little alternative but to reject convergentism as a position that lacks the support not only of considerations of general principles but also of the actual realities of our experience in the history of science.

This situation is critically important in our present context, for if later were indeed lesser, the ongoing cost escalation of scientific progress would not be so deeply problematic a factor. We would then simply forget about those very costly further purchases once a point is reached where their actual value becomes trivial. This is, however, most emphatically not the case in natural science, where economic limitations confront us with a real and serious obstacle.[5]

NOTES

1. Gunther S. Stent, *The Coming of the Golden Age* (Garden City, NY: Doubleday, 1969), 111-113. One notable exception to this view is that of the eminent Russian physicist Peter Kapitsa. After surveying various fundamental discoveries of the past he writes: "If we honestly extrapolate this curve we see it does not have any tendency towards saturation and that in the very near future many more such discoveries, which give us the possibility of increasing our control over nature and put new strength in our hands, will be made. Subjectively it seems that we know all there is to know about nature. However, when we read the works of scientists of the Newtonian era we see that they felt precisely the same. We can, therefore, be sure that further discoveries must still be made" (The Future Problems of Science" in M. Goldsmith and A. Mackay, eds., *The Science of Science* [London, 1964], pp. 102-113 [see 105-106]). This reflects standard, "party-line" thinking in the U.S.S.R.

2. See the author's *Peirce's Philosophy of Science* (Notre Dame and London: University of Notre Dame Press, 1978).

3. W. Stanley Jevons, *The Principles of Science*, 2nd ed. (London: Macmillan & Co., 1877), pp. 752-753.

4. The present critique of convergentism is thus very different from that of W. V. O. Quine. He argues that the idea of "convergence to a limit" is defined for numbers but not for theories, so that speaking of scientific change as issuing in a "convergence to a limit" is a misleading metaphor: "There is a faulty use of mathematical analogy in speaking of a limit of theories, since the notion of a limit depends on that of a 'nearer than,' which is defined for numbers and not for theories" (*Word and Object* [New York: John Wiley, 1960], p. 23). I am perfectly willing to apply the metaphor of substantial and insignificant differences to theories but am concerned to deny that, as a

matter of fact, the course of scientific theory innovation must eventually descend to the level of trivialities.

5. Some of the themes of this chapter are also addressed in Chapter 2 ("Scientific Progress as Nonconvergent") in the author's *Scientific Realism* (Dordrecht, Holland: D. Reidel, 1987).

Chapter Five

THE PROBLEM OF PREDICTING FUTURE SCIENCE

SYNOPSIS

(1) Our scientific "knowledge" about the world is fragile: the belief system of present-day natural science is—now and always—something transitory and impermanent. (2) It is difficult, indeed impossible, to predict the future of natural science, for we cannot forecast in detail even what the questions of future science will be—let alone the answers. (3) Present-day natural science cannot speak for future science. Viewed not in terms of its aims but in terms of its results, science is irremediably plastic: it is not something fixed, frozen, and unchanging but endlessly variable and protean, given to changing not only its opinions but its very form.

1. THE IMPERMANENCE OF THEORY IN NATURAL SCIENCE: WHY THEORIES FAIL

Nature herself is almost certainly not capricious; her laws presumably continue steadily in operation, ever persistent and unchanged. The same clearly cannot be said, however, for our *beliefs* about her laws, which are subject to ongoing revision. We have no way to get at nature's laws save via our ever-changing *theories* regarding their nature—this is all we ever have to work with. We can safely and unproblematically make the conditional prediction that *if* a generalization states a genuine law of nature, *then* the next century's phenomena will conform to it every bit as much as those of the last, but we can never in the prevailing condition of our information predict with unalloyed confidence that in the future people will still continue to regard those *theories* of ours as representing actual *laws* of nature.

Our understanding of the world's operations is based upon the theories of science—and especially physics, that most fundamental of natural sciences. But what can we predict regarding such theories themselves? Most securely—failure. Its epistemic instability—its change of mind about nature's modus operandi—is a fact of life *about* science that is as firmly established and confirmed as any of the facts *within* science itself. Any judgment we can make about the laws of nature— any model we can contrive regarding how things work in the world—is a matter of theoretical triangulation from the data at our disposal. We can never have assured confidence in the definitiveness of our database or in the adequacy of our theoretical exploitation of it; there is clearly good reason why no finite body of observations—however large—can ever settle decisively just what the laws of nature are.

So we come again to the opening chapter's theme of the processual nature of science. Much of our scientific knowledge is transitory. Theories arise, make their way to acceptance—be it gradually or dramatically—and eventually come to grief and fall out of favor. If historical experience teaches anything about the future, it teaches that we should expect many of our scientific theories of today to fall from favor by the year 3000, failing to survive without substantial amendment, modification, revision, or replacement. Not only is "the world" a thing of process but so is everything within it, specifically including "our world"—the world picture as we represent it in the science of the day.

Scientific progress brings in its wake not only new facts but also change of mind regarding the old ones; what we have in hand are not really certified "laws of nature" as such, but our *theories*—that is, *laws as we currently conceive them to be in the prevailing state of the scientific art*. Thus the realities of the scientific process—progress, change, discovery—mean that there is considerable opportunity for slippage.

The equilibrium between observation and theory that is realized by natural science at *any* given stage of its development is always an unstable one. The history of science shows all too clearly that many of our theories about the workings of nature have a finite lifespan. They come to be modified or replaced under various innovative pressures, in particular the enhancement of observational and experimental evidence (through improved techniques of experimentation, more powerful means of observation and detection, superior procedures for data processing, etc.). A state of the art of natural science is thus a human artifact that, like all other human creations, falls subject to the ravages of time. As fallibilists since C.S. Peirce's day have insisted, we must acknowledge an inability to attain a final and definitive truth in scientific matters. Historical experience and theoretical analysis alike indicate that stabilization with regard to theories is an unrealizable dream.

But why do theories fail? What aspect of reality accounts for this state of affairs? The answer ultimately lies in the fact that scientific theories are human products, and in the course of time all human contrivances fail. All human artifacts and constructions are fragile, transitory, and destined to ultimate collapse. Any human creation—be it a house, a dam, or a knowledge claim—is designed to function under certain known (or surmised) conditionsbut the processes of change that come with time always bring new, unforeseen, and unforeseeable circumstances to the fore. The best-laid productions of mice and men come to grief under the impact of the world's volatile contingencies.[1] This impermanence—this vulnerability to the pervasive dominion of chaos over the things of this world—holds just as much for our intellectual constructions as for our physical structures. It thereby predestines the ultimate failure of even our best predictive efforts. A scientific theory is a structure built to house and attuned to the needs of a certain body of experiences—fitted to the conditions of observation and information processing of a particular technological state of the art. As these conditions change, stresses and strains develop that destabilize the theory structure and lead to its eventual collapse. Changed social conditions destabilize social systems. Changed physical conditions destabilize structures. Changed experiential (i.e., experimental and observational) conditions—changed scientific technology, if you will—destabilize scientific theories. We learn by empirical inquiry about empirical inquiry, and one of the key things we learn is that at no actual stage does science yield a final and unchanging result. All the experience we can muster indicates that there is no justification for regarding our science as more than an inherently imperfect stage within an ongoing development. This fact has profound implications both for the predictions that we make through science and for those that we may make about it.

2. Difficulties in Predicting Future Science

The splendid dictum that "the past is a different country—they do things differently there" has much to be said for it, seeing that we generally cannot understand the past adequately in terms of the conceptions and presumption of the present. This is all the more drastically true of the human future—and the cognitive future in particular. After all, information about the thought world of the past is at any rate available—however laborious extracting it from the available data may prove to be. We do have in hand the material for forming a fairly plausible picture of what people were thinking in the past. We have, however, no comparable entryway into the intellectual future; we lack any viable pathway into the thought world of the years ahead. The past may be a different country, but the future is a terra incognita. Its science, its technology, its fads and fashions, etc., lie beyond our ken. We cannot begin to say what ideas will be at work here, though we know on general principles *that* they will differ from our own, but not *how*. Where our ideas cannot penetrate, we are ipso facto impotent to make any detailed predictions.

The best that we can do in matters of science and technology forecasting is to look toward those developments that are "in the pipeline" by looking to the reasonable extrapolation of character, orientation, and direction of the current state of the art—this is a powerful forecasting tool on the positive side of the issue. This approach rightly recognizes that in forecasting about science and technology we need to weigh heavily in our scale those directions of research that are currently vigorous. Pipeline flow analysis directs its attention to projects and investigations that are currently in progress or are applications or extensions of work that is presently underway.[2] Unlike trend extrapolation, which looks to the longer historical past, it focuses attention upon currently prominent work. One must never forget, however, the prospect of major innovation issuing from obscurity. The picture of Einstein toiling in the patent office in Bern should never be put altogether out of mind.

The long and short of it is that the future of science is an enigma for us. The landscape of natural science is ever-changing: innovation is the very name of the game. Not only do the theses and themes of science change but so do the very questions. Of course, once a body of science comes to be seen as something settled and firmly in hand, many issues become routine. Erstwhile major problems become mere reference questions—a matter of locating an answer within the body of preestablished, already available information that somewhere contains it. (The *mere* here is, to be sure, misleading in its downplaying of the formidable challenges that can arise in looking for needles in large haystacks.) In pioneering science, however, we face a very different situation. People may well wonder "what is the cause of X"—what causes cancer, say, or what produces the attraction of the lodestone for iron—in circumstances where the concepts needed to develop a workable answer still lie beyond their grasp. Indeed, most of the questions with which present-day science grapples could not even have been raised in the state of the art that prevailed a generation ago.

Commenting shortly after the publication of Frederick Soddy's speculations about atomic bombs in his 1930 book *Science and Life*,[3] Robert A. Millikan, a Nobel laureate in physics, wrote that "the new evidence born of further scientific study is to the effect that it is highly improbable that there is any appreciable amount of available subatomic energy to tap."[4] In science forecasting, the record of even the most qualified practitioners is poor, for people may well not even be able to conceive the explanatory mechanisms of which future science will make routine use. As one sagacious observer noted over a century ago:

> There is no necessity for supposing that the true explanation must be one which, with only our present experience, *we could* imagine. Among the natural agents with which we are acquainted, the vibrations of an elastic fluid may be the only one whose laws bear a close resemblance to those of light; but we cannot tell that there does not exist an unknown cause, other than an elastic ether diffused through space, yet producing effects identical in some respects with those which

would result from the undulations of such an ether. To assume that no such cause can exist, appears to me an extreme case of assumption without evidence.[5]

Since we cannot predict the answers to the presently open questions of natural science, we also cannot predict its future questions, for these questions will hinge upon those as yet unrealizable answers, since the questions of the future are engendered by the answers to those we have on hand. Accordingly, we cannot predict science's solutions to its problems because we cannot even predict in detail just what these problems will be.

In scientific inquiry, as in other areas of human affairs, major upheavals can occur in a manner that is sudden, unanticipated, and often unwelcome. Major breakthroughs often result from research projects that have very different ends in view. Louis Pasteur's discovery of the protective efficacy of inoculation with weakened disease strains affords a striking example. While studying chicken cholera, Pasteur accidentally inoculated a group of chickens with a weak culture. The chickens became ill, but, instead of dying, recovered. Pasteur later reinoculated these chickens with fresh culture—one strong enough to kill an ordinary chicken. To Pasteur's surprise, the chickens remained healthy. Pasteur then shifted his attention to this interesting phenomenon, and a productive new line of investigation opened up. In empirical inquiry, we generally cannot tell in advance what further questions will be engendered by our endeavors to answer those on hand. New scientific questions arise from answers we give to previous ones, and thus the issues of future science simply lie beyond our present horizons.

It would, after all, be quite unreasonable to expect detailed prognostications about the particular *content* of scientific discoveries. We know—or at any rate can safely predict—*that* future science will make major discoveries (both theoretical and observational/phenomenological) in the next century, but we cannot say *what* they are and *how* they will be made (since otherwise we could proceed to make them here and now). We could not possibly predict now the substantive content of our future discoveries—those that result from our future cognitive choices. For to do so would be to transform them into present discoveries that, by hypothesis, they just are not.[6] In matters of scientific importance, then, we must be prepared for surprises.

3. IN NATURAL SCIENCE, THE PRESENT CANNOT SPEAK FOR THE FUTURE

It is a key fact of life that ongoing progress in scientific inquiry is a process of *conceptual* innovation that always places certain developments outside the cognitive horizons of earlier workers because the very concepts operative in their characterization become available only in the course of scientific discovery itself. (Short of learning our science from the ground up, Aristotle could have made nothing of modern genetics.) What one scientific generation sees as a natural kind, a later one disassembles into a variety of different species. We have as yet no inkling of the concept mechanisms that later scientific eras will make use of. The major discoveries of later stages are ones that the workers of a substantially earlier period (however clever) not only failed to make but could not even have *understood*, because the requisite concepts were simply not available to them. Newton could not have predicted findings in quantum theory any more than he could have predicted the outcome of American presidential elections. One can make predictions only about what one is cognizant of, takes note of, deems worthy of consideration. Thus, it is effectively impossible to predict not only the answers but even the questions that lie on the agenda of future science, for new questions in science always arise out of the answers we give to old ones, and the answers to these questions involve conceptual innovations. We cannot now predict the future states of scientific knowledge in detail because we do not yet have at our disposal the very concepts in which the issues will be posed.

In cognitive forecasting, it is the errors of omission—our blind spots, as it were—that present the most serious threat, for the fact is that we cannot substantially anticipate the evolution of knowledge. Given past experience, we can feel reasonably secure when we say *that* science will resolve various problems in the future, but *how* it will do so is bound to be a mystery.[7]

With respect to the major substantive issues of future natural science, we must be prepared for the unexpected. If there was one thing of which the science of the first half of the seventeenth century was confident, it was that natural processes were based on contact interaction and that there can be no such thing as action at a distance. Newtonian gravitation burst upon this scene like a bombshell. Newton's supporters simply stonewalled. Roger Cotes explicitly denied there was a problem, arguing (in his preface to the second edition of Newton's *Principia*) that nature was *generally* unintelligible, so that the unintelligibility of forces acting without contact was nothing specifically worrisome. However unpalatable Cotes's position may seem as a precept for science (given that making nature's workings understandable is, after all, one of the aims of the enterprise), there is something to be said for it—not, to be sure, as science but as meta-science, for we cannot hold the science of tomorrow bound to the standards of intelligibility espoused by the science of today. The cognitive future is inaccessible to even the ablest of present-day workers. After Pasteur had shown that bacteria could come only from pre-existing bacteria, Darwin wrote that "it is mere rubbish thinking of the origin of life; one might as well think of the origin of matter."[8] One might indeed!

The inherent impredictability of scientific developments—the fact that inferences from one state of science to another are generally precarious—bears importantly upon the issue of future science and its limits, for in matters of substance, *present-day science cannot speak for future science*: it is in principle impossible to make any secure inferences from the substance of natural science at one time to its substance at a significantly different time. The prospect of future scientific revolutions can never be precluded. We cannot say with unblinking confidence what sorts of resources and conceptions the science of the future will or will not use. Given that it is effectively impossible to predict the details of what future science will accomplish, it is no less impossible to predict in detail what future science will *not* accomplish. We can never confidently put this or that range of issues outside "the limits of science," because we cannot discern the shape and substance of future discoveries with sufficient clarity to be able to say with any assurance what it can and cannot do. Any attempt to set limits to natural science—any advance specification of what it can and cannot do by way of handling problems and solving questions—is destined to come to grief. Our present-day natural science cannot speak for the substance of future science. Viewed not in terms of its *aims* but in terms of its *results*, science is inescapably plastic: it is not something fixed, frozen, and unchanging but endlessly variable and protean—given to changing not only its opinions but its very form.

An apparent violation of the rule that present science cannot bind future science is afforded by John von Neumann's attempt to demonstrate that all future theories of subatomic phenomena—and thus all *future* theories—will have to contain an analogue of Heisenberg's uncertainty principle if they are to account for the data explained by present theory. Complete predictability at the subatomic level, he argued, was thus exiled from science, but the "demonstration" proposed by von Neumann in 1932 places a substantial burden on potentially changeable details of presently accepted theory.[9] The fact remains that we cannot preclude fundamental innovation in science: present theory cannot delimit the potential of future discovery. In natural science we cannot erect our structures with a solidity that defies demolition and reconstruction. Even if the existing body of "knowledge" does confidently and unqualifiedly support a certain position, this circumstance can never be viewed as absolutely final.

Not only can one never claim with confidence that the science of tomorrow will not resolve the issues that the science of today sees as intractable, but one can never be sure that the science of tomorrow will not endorse principles that the science of today rejects. This is why it is infinitely risky to speak of this or that explanatory resource (action at a distance, stochastic processes, mesmerism, etc.) as inherently unscientific. Even if X lies outside the range of science as we nowadays construe it, it by no means follows that X lies outside science as such. We must recognize the commonplace phenomenon that the science of the day almost always manages to do what the science of an earlier day deemed infeasible to the point of absurdity (split the atom, abolish parity, etc.). With natural science, the substance of the future inevitably lies beyond our present grasp. No one is in a position to delineate here and now what the science of the future can and cannot achieve. Here we can set no *a priori* restrictions but, rather, have to be flexible.

To be sure, when it comes to considering the *volume* of discovery in future science—rather than its *content* —reasonable prediction does seem to be possible. This is clear from the previous chapter's discussion of the nonasymptotic character of scientific progress, and the next chapter will extend this range of consideration. Of course, this sort of information tells us only about the *structure* of future science and not about its *substance*. The fact is that while we can predict *that* future science will be more complex than ours, we cannot foresee *how* it will be so. We know that the adequate exposition of the physics or chemistry (etc.) of the future will take up many more pages than ours, but we do not have a clue as to what it is that those pages will say. We know that the science of the future will differ comparably from ours but cannot say exactly where and how this difference will come into play. Ironically, natural science—our most powerful predictive tool—is itself unpredictable as concerns matters of substance.[10]

NOTES

1. See Charles Perrow, *Normal Accidents* (New York: Basic Books, 1989).

2. The science forcasts produced in the late 1970s by the National Science Foundation were primarily a large-scale exercise in pipeline flow analysis, with a bit of trend extrapolation mixed in. See, for example, *The Five Year Outlook: Problems, Opportunities, and Constants in Science and Technology*, 2 vols. (Washington DC: National Science Foundation, 1980).

3. Frederick Soddy, *Science and Life* (New York: E. P. Dutton, 1930).

4. Quoted in *Daedalus*, vol. 107 (1978), p. 24.

5. Baden Powell, *Essays on the Spirit of the Inductive Philosophy* (London: Gregg, 1855), p. 23.

6. As one commentator has wisely written: "But prediction in the field of pure science is another matter. The scientist sets forth over an uncharted sea and the scribe, left behind on the dock, is asked what he may find at the other side of the waters. If the scribe knew, the scientist would not have to make his voyage" (anonymous, "The

Future as Suggested by Developments of the Past Seventy-Five Years," *Scientific American*, vol. 123 [1920], p. 321). The role of unforeseeable innovations in science forms a key part of Popper's case against the impredictability of man's social affairs—given that new science engenders new technologies, which in turn make for new modes of social organization. (See K. R. Popper, *The Poverty of Historicism* [London: Routledge & Kegan Paul, 1957], pp. vi and passim. The impredictability of revolutionary changes in science also figures centrally in W. B. Gallie's "The Limits of Prediction" in S. Körner, ed., *Observation and Interpretation* [New York: Oxford University Press, 1957]). Gallie's argumentation is weakened, however, by a failure to distinguish between the generic fact of future discovery in a certain domain and its specific nature. See also Peter Urbach, "Is Any of Popper's Arguments Against Historicism Valid?" *British Journal for the Philosophy of Science*, vol. 29 (1978), pp. 117-30 (see pp. 128-29), whose deliberations seem (to this writer) to skirt the key issues. A judicious and sympathetic treatment is given in Alex Rosenberg, "Scientific Innovation and the Limits of Social Scientific Prediction," *Synthese*, vol. 97 (1993), pp. 161-81. On the present issue Rosenberg cites the anecdote of the musician who answered the question "Where is jazz heading" with the response: "If I knew that, I'd be there already" (*op. cit.*, p. 167).

7. This holds for technology as well through the principle of "equivalent invention" described by S. C. Gilfillin in 1939. (For various references see S. Colum Gilfillan, "A Sociologist Looks at Technical Prediction," in James R. Wright, ed., *Technical Forecasting for Industry and Government* (Englewood Cliffs, NJ: McGraw-Hill, 1968). Instancing the problem of flying aircraft in fog, Gilfillan notes that some dozen different ways of addressing it were under consideration in the 1930s, but which method would prove most successful and thus prevail was imponderable. That the problem would be resolved could be predicted with confidence, but *how* was unforeseeable.

8. Quoted in Philip Handler, ed., *Biology and the Future of Man* (Oxford: Clarendon Press, 1970), p. 165.

9. See also the criticisms of his argument in David Bohm, *Causality and Chance in Modern Physics* (London: Routledge and Kegan Paul, 1967), pp. 95-96. Bohm's hidden variable theory provides a counterexample to von Neumann's argumentation—though it is one that many physicists are reluctant to adopt.

10. Some of the ideas of the preceding discussion are also discussed in the author's *The Limits of Science* (Berkeley, Los Angeles, London: University of California Press, 1984). The book is also available in translation: German transl., *Die Grenzen der Wissenschaft* (Stuttgart: Reclam, 1984), Spanish transl., *Los Límites de le ciencia* (Madrid: Editorial Tecnos, 1994), Italian transl., *I Limiti della scientia* (Rome: Armando Editore, 1990).

Chapter Six

THE LAW OF LOGARITHMIC RETURNS

SYNOPSIS

(1) Scientific knowledge does not correlate with the brute volume of scientific information, but only with its logarithm. (2) The rationale of this circumstance roots in the way in which significant information always lies obscured amid a fog of insignificance. (3) In consequence, the progress of knowledge involves ever-escalating demands, a circumstance that bears in a fundamental and ominous way on the issue of the growth of scientific knowledge over time. (4) To be sure, the law of logarithmic returns pivots in a critical way on assessing the quality of the work at issue, but of course anything deserving of the name of scientific knowledge will be information of the highest quality level. Thus there is indeed a prospect of making some economically based predictions about the volume of future scientific innovation, despite our inability to foresee its content.

1. THE LAW OF LOGARITHMIC RETURNS

Certain evaluative distinctions and classifications have played a pivotal role in epistemology and the theory of cognition, for example, the familiar distinctions between the true and the false and between the well evidentiated (probable) and the evidentially counter indicated (improbable). However, the portentous distinction between real knowledge and mere information has been generally neglected. This is eminently unfortunate. After all, items of information were not created equal. Some claims will, even if true, be insignificant and make little or no impact upon the larger cognitive scheme of things. (What difference does my personal preference for pink over purple make to anything?) Other truths will be distinctly important—the theory of relativity for example, or Avogadro's law in chemistry. Few issues regarding cognition can rival this distinction in point of its far-reaching bearing upon matters of discovery, learning, understanding, and insight—alike in scientific and in everyday contexts, and so it will be this distinction that concerns us here.

Enroute to knowledge we must begin with information. How can one measure the volume of information generated in a field of suitable or scholarly inquiry? Various ways suggest themselves. The size of the literature of a field (as measured by the sheer bulk of publication in it) affords one possible measure, and there are various other possibilities as well. One can also proceed by way of inputs rather than outputs, measuring, for example, the number of workdays that investigators devote to the topics at issue or the amount of resource investment in the relevant information-engendering technology. So much for the quantitative assessment of *information*. But what about *knowledge*?

By "knowledge" we shall here understand *putative* knowledge that is not necessarily correct but merely represents a conscientiously contrived best estimate of what the truth of the matter actually is.[1] This is not, however, all there is to it. Information is simply a collection of (supposedly correct) beliefs or assertions, while knowledge, by contrast, is something more select, more deeply issue resolving. Like all usable information, it must be duly evidentiated and at least *presumably* correct, but there is also an additionally evaluative aspect, seeing that knowledge is a matter of *important* information: information that is *significantly* informative. Not every insignificant smidgeon of information constitutes knowledge, and the person whose body of information consists of utter trivia really knows virtually nothing.

To provide a simple illustration for this matter of significance, let us   suppose an object-descriptive color taxonomy—for the sake of example, an oversimple one based merely on Blue, Red, and Other. Then that single item of *knowledge* represented by "knowing the color" of an object—viz., that it is red—is bound up with many different items of (correct) *information* on the subject (that it is not Blue, is rather similar to some shades of Other, etc.). As such information proliferates, we confront a situation of redundancy and diminished productiveness. Any knowable fact is always potentially surrounded by a vast penumbral cloud of relevant information, and as our information grows to be ever more extensive, those really *significant* facts become more difficult to discern. Knowledge certainly increases with information, but at a far less than proportional rate.

One instructive way of measuring the volume of available information is via the opportunities for placement in a framework for describing or classifying the features of things. With n concepts, you can make n^2 two-concept combinations. With m facts, you can project m^2 fact-connecting juxtapositions, in each of which some sort of characteristic relationship is at issue.[2] A single step in the advancement of knowledge is here accompanied by a massive increase in the proliferation of information. Extending the previous example, let us also contemplate *shapes* in addition to *colors*, again supposing only three of them: Rectangular, Circular, Other. Now when we *combine* color and shape there will be $9 = 3 \times 3$ possibilities in the resultant (cross-) classification. So with that complex, dual-aspect piece of knowledge (color + shape) we also launch into a vastly amplified (i.e., multiplied) information spectrum over that increased classification space. In moving cognitively from n to $n + 1$ cognitive parameters we enlarge our knowledge additively but expand our information field multiplicatively.

If a control mechanism can handle three items—be it in physical or in cognitive management—then we can lift ourselves to higher levels of capacity by hierarchical layering. First, we can, by hypothesis, manage three base-level items; next, we can, by grouping three of these into a first-level complex, manage nine base-level items; and thereupon by grouping three of these first-level complexes into a second-level complex we can manage 27 base-level items. Hierarchical grouping is thus clearly a pathway to enhanced managerial capacity. In cognitive contexts of information management, however, it is clear that a few items of high-level information (= knowledge) can and will correspond to a vast range of low-level information

(= mere truths). Knowing (in our illustration) where we stand within each of those three levels—that is, having three pieces of *knowledge*—will position us in a vastly greater information space (viz., one of 27 compartments). This situation is typical. The relational structure of the domain means that a small range of knowledge (by way of specifically high-grade information) can always serve to position one cognitively within a vastly greater range of low-level information.

It is instructive to view this idea from a different point of view. Knowledge commonly develops via distinctions (A vs. non-A) that are introduced with ever-greater elaboration to address the problems and difficulties that one encounters with less sophisticated approaches. A situation obtains that is analogous to the game of "20 Questions" with an exponentially exfoliating possibility space being traced out stepwise ($2, 4, 8, 16, \ldots, 2^n, \ldots$). With n descriptors one can specify for 2^n potential descriptions that specify exactly how, over all, a given object may be characterized. When we add a new descriptor, we increase by one additional unit the amount of knowledge but double the amount of available information. The *information* at hand grows with 2^n, but the knowledge acquired merely with n. The cognitive exploitation of information is a matter of dramatically diminishing returns.

Consider an illustration of a somewhat different nature. The yield of knowledge from information afforded by legibility-impaired manuscripts and papyri and inscriptions in classical paleography provides an instructive instance. If we can decipher 70 percent of the letters in such a manuscript, we can reconstruct the phrase at issue. If we can make out 70 percent of the phases we can pretty well figure out the sentences. If we can read 70 percent of its sentences, we can understand the message of the whole text. So some one-third of the letters suffice to carry the whole message. A vast load of information stands coordinate with a modest body of knowledge. From the standpoint of knowledge, information is highly redundant, albeit unavoidably so.

A helpful perspective on this situation comes to view through the idea of "noise," considering that expanding bodies of information encompass so much cognition-impeding redundancy and unhelpful irrelevancy that it takes successive many-fold increases in information to effect successive fixed-size increases in actual knowledge. It is supposed that internal "noise" or variability in the information system is such that the increment needed to effect a cognitive advance of a certain fixed size or significance is proportional to the magnitude of the starting position as a whole.

Consider now the result of combining two ideas, namely,

(1) knowledge is distinguished from mere information as such by its significance. In fact: *Knowledge is simply particularly significant information*—information whose significance exceeds some threshold level (say q). (In principle there is room for variation here as one sets the quality level of entry qualification and the domain higher or lower.)

(2) The significance of *additional* information is determined by its impact upon *preexisting* information.

Significance in this sense is a matter of the relative (percentage-wise) increase that the new acquisitions effect upon the body of preexisting information (I), which may—to reiterate—be estimated in the first instance by the sheer volume of the relevant body of information: the literature of the subject, as it were. Accordingly: *The significance of incrementally new information can be measured by the ratio of the increment of new information to the volume of information already in hand*: $\Delta I / I$.

Putting the ideas of these two principles together, we have it that a new item of actual knowledge is one for which the ratio of information increments to preexisting information exceeds a fixed threshold-indicative quantity q:

$$\frac{\Delta I}{I} \geq q$$

Thus knowledge-constituting *significant* information is determined through the proportional extent of the change effected by a new item in the preexisting situation (independently of what that preexisting situation is).

On the basis of these considerations, it follows that the cumulative total amount of knowledge (K) encompassed in an overall body of information of size I is given by the *logarithm* of I. This is so because we have:

$$K = \int \frac{dI}{I} = \log I + \text{const} = \log cI$$

where K represents the volume of actual *knowledge* that can be extracted from a body of bare *information*

I,[3] with the constant at issue open to treatment as a unit-determinative parameter of the measuring scale, so that the equation at issue can be simplified without loss to $K = \log I$.[4]   In milking additional information for significant insights it is generally the *proportion* of the increase that matters: its percentage rather than its brute amount. We accordingly arrive at the *Law of Logarithmic Returns* governing the extraction of significant *knowledge* from bodies of mere *information*.

It should thus come as no surprise that knowledge coordinates with information in multiplicative layers. With texts, we have the familiar stratification: article/chapter, book, library, library system, or pictographically: sign, scene (= ordered collection of signs), cartoon (= ordered collection of scenes to make it a story). In such layering, we have successive levels of complexity corresponding to successive levels of informational combining combinations, proportional with n, n^2, n^3, etc. The logarithm of the levels—$\log n$, $2\log n$, $3\log n$, etc.—reflects the amount of "knowledge" that is available through the information we obtain about the state of affairs prevailing at each level. This is something that increases only one step at a time, despite the exponential increase in information.

The general purport of such a Law of Logarithmic Returns as regards expanding information is clear enough. Letting $K(I)$ represent the quantity of knowledge inherent in a body of information I, we begin with our fundamental relationship: $K(I) = \log I$. On the basis of this fundamental relationship, the knowledge of a two-sector domain increases additively notwithstanding a multiplicative explosion in the amount of information that is now upon the scene. If the field

(F) under consideration consists of two subfields (F_1 and F_2), then because of the cross connections obtaining within the information domains at issue the overall information complex will take on a multiplicative size:

$$I \ = \ \inf{(F)} \ = \ \inf{(F_1)} \ \text{x} \ \inf{(F_2)} \ = I_1 \ \text{x} \ I_2$$

With compilation, *information* is multiplied, but in view of the indicated logarithmic relationship, the *knowledge* associated with the body of compound information I will stand at:

$$K(I) = \log I = \log{(I_1 \ \text{x} \ I_2)} = \log I_1 + \log I_2 \ = \ K(I_1) + K(I_2)$$

The knowledge obtained by compiling two information domains (subfields) into an overall aggregate will (as one would expect) consist simply in *adding* the two bodies of knowledge at issue. Where compilation increases *information* by multiplicative leaps and bounds, the increase in *knowledge* is merely additive.

2. The Rationale and Implications of the Law

The Law of Logarithmic Returns presents us with an epistemological analogue of the old Weber-Fechner law of psychophysics, asserting that inputs of geometrically increasing magnitude are required to yield *perceptual* outputs of arithmetically increasing intensity. The presently contemplated law envisions a parallelism of perception and conception in this regard. It stipulates that (informational) inputs of geometrically increasing magnitude are needed to provide for (cognitive and thus) *conceptual* outputs of arithmetically increasing magnitude. (In the search for meaningful patterns, additional data points make a contribution of rapidly diminishing value.)

It is not too difficult to come by a plausible explanation for the sort of information/knowledge relationship that is represented by $K = \log I$. The principal reason for such a K/I imbalance may lie in the efficiency of intelligence in securing a view of the modus operandi of a world whose law structure is comparatively simple. Here one can learn a disproportionate amount of general fact from a modest amount of information. (Note that whenever an infinite series of 0's and 1's, as per 01010101 . . ., is generated—as this series indeed is—by a relatively *simple* law, then this circumstance can be gleaned from a comparatively short initial segment of this series.) In rational inquiry we try the simple solutions first, and only if and when they cease to work—when they are ruled out by further findings (by some further influx of coordinating information)—do we move on to the more complex. Things go along smoothly until an oversimple solution becomes destabilized by enlarged experience. We get by with the comparatively simpler options until the expanding information about the world's modus operandi made possible by enhanced new means of observation and experimentation demands otherwise. With the expansion of knowledge, however, new accessions set ever-increasing demands. At bottom, then, there are thus two closely interrelated reasons for that K/I disparity:

i. Where order exists in the world, intelligence is rather efficient in finding it.

ii. If the world were not orderly—were not amenable to the probes of intelligence—then intelligent beings would not and could not have emerged in it through evolutionary mechanisms.

The implications for cognitive progress of this disparity between knowledge and mere information are not difficult to discern. Nature imposes increasing resistance barriers to intellectual as to physical penetration. Consider the analogy of extracting air for creating a vacuum. The first 90 percent comes out rather easily. The next 9 percent is effectively as difficult to extract as all that went before. The next .9 percent is proportionally just as difficult, and so on. Each successive order-of-magnitude step involves a massive cost for lesser progress; each successive fixed-size investment of effort yields a substantially diminished return. Intellectual progress is exactly the same: when we extract actual *knowledge* (i.e., high-grade, nature-descriptively significant information) from mere information of the routine, common "garden variety," the same sort of quantity/quality relationship obtained. Initially a sizable proportion of the available knowledge is high grade—but as we press further, this proportion of what is cognitively significant gets ever smaller. To double knowledge we must quadruple information. As science progress, the important discoveries that represent real increases in knowledge are surrounded by an ever-increasing penumbra of mere items of information. (The mathematical literature of the day yields an annual crop of over 200,000 new theorems.[5])

This state of affairs clearly has the most far-reaching implications for the progress of knowledge. This is illustrated in Display 1. There is an immense K/I imbalance:

Display 1
THE STRUCTURE OF THE KNOWLEDGE/
INFORMATION RELATION

I	cI (with $c = .1$)	log cI (=K)	$\Delta(K)$	K as % of I
100	10	1	—	1.0
1,000	100	2	1	.2
10,000	1000	3	1	.03
100,000	10,000	4	1	.004

With continuously increasing information I, the corresponding increase in knowledge K, shrinks markedly. To increase knowledge by equal steps, we must amplify information by successive orders of magnitude.

Extracting knowledge from information thus requires ever-greater effort, for the greater a body of information, the larger the *patterns of order* that can potentially obtain and the greater the effort needed to bring particular orderings to light. With two-place combinations of the letters A and B (yielding the four pairs *AA, AB, BA*, and *BB*) we have only two possible patterns of order—namely, "The same letter all the way through" (*AA* and *BB*), and "Alternating letters" (*AB* and *BA*). As we add more letters, the possibilities proliferate massively.

This situation is reflected in Max Planck's appraisal of the problems of scientific progress. He wrote that *"with every advance [in science] the difficulty of the task is increased; ever larger demands are made on the achievements of researchers,* and the need for a suitable division of labor becomes more pressing."[6] The Law of Logarithmic Returns at once characterizes and explains this circumstance of why substantial findings are thicker on the informational ground in the earlier phase of a new discipline and become ever more attenuated in the natural course of progress.

3. THE GROWTH OF KNOWLEDGE

The Law of Logarithmic Returns also connects up with—and is interestingly illustrated by—the idea of a cognitive lifespan as expounded by the sagacious Edward Gibbon in his *Memoirs of My Life*: "The proportion of a part to the whole is the only standard by which we can measure the length of our existence. At the age of twenty, one year is a tenth perhaps of the time which has elapsed within our consciousness and memory; at the age of fifty it is no more than a fortieth, and this relative value continues to decrease till the last sands are shaken out [of the hour-glass measure of our lifespan] by the hand of death."[7] On this basis, knowledge development is a matter of adding a given percentage increment of what has gone before. Thus fresh experience superadds its additional increment ΔE to the preexisting total E in such a way that its effective import is measured by the proportion that movement bears to the total: $\Delta E / E$. Cumulatively, we of course have it that this comes to the logarithm of E: $\int \Delta E / E = \log E$. On such an approach, an increment to one's lifetime has a *cognitive* value determined on strict analogy with Daniel Bernoulli's famous proposal to measure the *utility* value of incremental economic resources by means of a logarithmic yardstick.[8]

The Law of Logarithmic Returns clearly has substantial implications for the *rate* of scientific progress.[9] In particular, it stands coordinate with a principle of a swift and steady decline in the comparative cognitive yield of additions to our body of mere information. With the enhancement of scientific technology, the size and complexity of this body of data inevitably grow, expanding on quantity and diversifying in kind. Technological progress constantly enlarges the window through which we look out upon nature's parametric space. In developing natural science, we use this window to scrutinize parametric space, continually augmenting our database and then generalizing upon what we see. What we have here is not a homogeneous landscape, where once we have seen one sector we have seen it all, and where theory projections from lesser data generally remain in place when further data comes our way. Historical experience shows that there is every reason to expect that our ideas about nature are subject to constant radical changes as we enhance the means of broadening our database. Of course the expansion of knowledge proceeds at a far slower rate than the increase of bare information.

It is illuminating to look at the implications of this state of affairs in an historical perspective. The salient point is that the growth of knowledge over time involves ever-escalating demands. Progress is always possible—there are no absolute limits, but increments of the same cognitive magnitude have to be obtained at an ever-increased price in point of information development and thus of resource commitment as well. Here our basic formula that knowledge stands proportionate to the logarithm of information ($K \approx \log I$) means the increase of knowledge over time stands to the increase of information in a proportion fixed by the *inverse* of the volume of already available information:

$$\frac{d}{dt} K \approx \frac{d}{dt} \log I \approx \frac{1}{I} \frac{d}{dt} I$$

The more knowledge we already have in hand, the slower (by very rapid decline) will be the rate at which knowledge grows with newly acquired information. The larger the body of information we have, the smaller will be the proportion of this information that represents real knowledge.

In developmental perspective, there is good reason to suppose that our body of bare *information* will increase more or less in proportion with our resource investment in information gathering, but then, if this investment grows exponentially over time (as has historically been the case in the recent period), we shall in consequence have it that

$$I(t) \approx c^t \text{ and correspondingly also } \frac{d}{dt} I(t) \approx c^t$$

and accordingly

$$K(t) \approx \log I(t) \approx \log c^t = t$$

and consequently:

$$\frac{d}{dt} K(t) = \text{constant}$$

Exponential growth in I is coordinated with linear growth in K.

On this basis, it is not all that difficult to find empirical substantiation of our law of logarithmic returns $(K \approx \log I)$. The phenomenology is that while information has increased exponentially in the past (as shown by the exponential increase in journals, scientists, and outlays for the instrumentalities of research), real knowledge has expanded only linearly. Throughout modern times the number of scientists has been increasing at roughly 5 percent per annum.[10] Thus well over 80 percent of ever-existing scientists (in even the oldest specialties, e.g., mathematics, physics, and medicine) are alive and active at the present time. Scientific information has also been growing at the (reasonably constant) exponential rate over the past several centuries, so as to produce a veritable flood of scientific literature in our time. The *Physical Review* is now divided into six parts, each of which is larger than the whole journal was a decade or so ago. The total volume of scientific publication is truly staggering. It is reliably estimated that, from the start, well over 10 million scientific papers have been published and that by the mid-1970s some 30,000 journals were publishing some 600,000 new papers each year. It is readily documented that the number of books, journals, and journal papers has been increasing at an exponential rate over the recent period.[11] To be precise, the printed literature of science has been increasing at some 5 percent per annum throughout the last two centuries, to yield an exponential growth rate with a doubling time of about 15 years and an order-of-magnitude increase roughly every half century. Much the same story holds for the recruitment of people and the expenditures on equipment—that is, for resource commitment in general.

During this period the progress of science per se —the progress of authentic scientific knowledge as measured in the sort of first-rate discoveries of the highest level of significance—has progressed at a more or less linear rate. There is much evidence to this claim. It suffices to consider the size of encyclopedias and synoptic textbooks, or again, the number of awards given for really big contributions (Nobel Prizes, academy memberships, honorary degrees), or the expansion of the classificatory taxonomy of branches of science and problem areas of inquiry.[12] All of these measures of preeminently *cognitive* contribution conjoin to indicate that there has in fact been a comparative constancy in year-to-year progress, the exponential growth of the scientific enterprise notwithstanding.

4. The Centrality of Quality and Its Implications

The preceding deliberations also serve to indicate, however, why the question of the rate of scientific progress is somewhat tricky. This whole issue will turn rather delicately on fundamentally *evaluative* considerations. Thus consider the past once more. At the crudest—but also most basic—level, where progress is measured simply by the growth of the scientific literature, there has for centuries been astonishingly swift and sure progress: an exponential growth with a doubling time of roughly 15 years. At the more sophisticated and demanding level of high quality results of a suitably "important" character, there has also been exponential growth, but only at the pace of a far longer doubling time, perhaps 30 years, approximating the reduplication with each successive generation envisaged by Henry Adams at the turn of this century.[13] Finally, at the maximally exacting level of the really crucial insights that fundamentally enhance our picture of nature, our analysis has it that science has been maintaining a merely constant pace of progress.

It is this last consideration that is crucial for present purposes and brings us back to the point made at the very outset. What we have been dealing with here is an essentially seismological standard of importance. It is based on the question, "If the thesis at issue were abrogated or abandoned, how large would the ramifications and implications of this circumstance be? How extensive would be the shocks and tremors felt throughout the whole range of what we (presumably) know?" What is at issue is exactly a kind of cognitive Richter scale based on the idea of successive orders of magnitude of impact. The crucial determinative factor for increasing importance is the extent of seismic disturbance of the cognitive terrain. Would we have to abandon and/or rewrite the entire textbook, or a whole chapter, or a section, or a paragraph, or a sentence, or a mere footnote?

So, while the question of the rate of scientific progress does indeed involve the somewhat delicate issue of setting of an evaluative standard, our stance here can be rough and ready—and justifiably so because the details pretty much wash out. *Science viewed as a cognitive discipline*—a body of knowledge whose task is the unfolding of a rational account of the modus operandi of nature—stands correlative with its accession of really major discoveries: the seismically significant, cartography-revising insights into nature. Here, progress has historically been sure and steady—but essentially linear. However—and this perspective lays the foundation for our present analysis —the historical situation has been one of a *constant* progress of science as a *cognitive discipline* notwithstanding its *exponential* growth as a *productive enterprise* (as measured in terms of resources, money, manpower, publications, etc).[14]

If we look at the cognitive situation of science in its quantitative aspect, the Law of Logarithmic Returns pretty much says it all. In the struggle to achieve cognitive mastery over nature, we have been confronting an enterprise of ever-escalating demands, with the exponential growth in the *enterprise* associated with a merely linear growth in the *discipline*. This situation regarding the past has important implications for the future, for while one cannot—as we have seen—hope to predict the *content* of future science, the law does actually put us into a position to make plausible estimates about its *volume*.[15]

NOTES

1. There is thus no suggestion of cumulativeness here. Clearly, knowledge, like population, grows not just by additions (births) but

also by abandonment (deaths), but with knowledge, in contrast to population, the concepts of displacement and replacement play a more crucial role. Knowledge never dies off without leaving progeny: in this domain you cannot eliminate something with nothing.

2. Note that a self-concatenated concept or fact is still a concept or fact, even as a self-mixed color is still a color.

3. In information theory, *entropy* is the measure of the information conveyed by a message and is measured by $k \log M$, where M is the number of structurally equivalent messages constructable by the available sorts of symbols. By extension, the $\log I$ measure of the knowledge contained in a given body of information might accordingly be designated as the *enentropy*. Either way, the concept at issue measures informative actuality in relation to informative possibility, for there are two types of informative possibilities: (1) structural/syntactical as dealt with in classical information theory, and (2) hermeneutic/semantical (i.e., genuinely meaning oriented) as dealt with in the present theory.

4. Some writers have suggested that the subcategory of significant information included in an overall body of crude data of size I should be measured by I^k (for some suitably adjusted value $0 < k < 1$)—for example, by the "Rousseau's Law" standard of the square root of I. (For details see Chapter VI of the author's *Scientific Progress* [Oxford: Blackwell, 1978].). Now, since $\log I^k = k \log I'$ which is proportional to $\log I$, the specification of *this* sort of quality level for information would again lead to a $K \approx \log I$ relationship.

5. See Stanislaw M. Ulam, *Adventures of a Mathematician* (New York: Scribner, 1976).

6. Max Planck, *Vorträge und Erinnerungen*, 5th ed. (Stuttgart: S. Hirzel, 1949), 376; italics added. Shrewd insights seldom go unanticipated, so it is not surprising that other theorists should be able to contest claims to Planck's priority here. C. S. Peirce is particularly noteworthy in this connection.

7. Edward Gibbon, *Memoirs of My Life* (Harmondworth, UK: Penguin, 1984), 63.

8. Gibbon's "law of learning" thus means that a body of experience that grows linearly over time yields a merely logarithmic growth in *cognitive* age. Thus a youngster of ten years has attained only one-eighth of his or her chronological expected lifespan but has already passed the halfway mark of his or her cognitive expected lifespan.

9. It might be asked: "Why should a mere accretion in scientific 'information'—in mere belief—be taken to constitute *progress*, seeing that those later beliefs are not necessarily *true* (even as the earlier ones were not)?" The answer is that when we are dealing with *rational* people, their later help will in any case be better *substantiated* —that they are "improvements" on the earlier ones by way of the elimination of shortcomings. For a detailed consideration of the relevant issues, see the author's *Scientific Realism* (Dordrecht, Holland: D. Reidel, 1987).

10. Derek J. deSolla Price, *Little Science, Big Science* (New York: Columbia University Press, 1963), pp. 6-8.

11. For the statistical situation in science see Derek J. deSolla Price, *Science Since Babylon* (New Haven: Yale University Press, 1961; 2nd ed. 1975). See Chap. 8, "Diseases of Science." Further detail is given in Price's *Little Science, Big Science* (op. cit.).

12. The data here are set out in the author's *Scientific Progress* (Oxford: 1978).

13. Henry Brooks Adams, *The Education of Henry Adams* (Boston: Haughton-Mifflin, 1918).

14. To be sure, we are caught up here in the usual cyclic pattern of all hypothetico-deductive reasoning. In addition to *explaining* the various phenomena we have been canvassing, that projected K/I relationship is in turn *substantiated* by them. This is not a vicious circularity but simply a matter of the systemic coherence that lies at the basis of inductive reasonings. Of course the crux is that there also be some predictive power, which is exactly what our discussion of deceleration is designed to exhibit.

15. Some of the themes of this chapter were also addressed in the author's *Scientific Progress* (Oxford: Blackwell, 1978). This book is also available in translation: German transl., *Wissenschaftlicher Fortschritt* (Berlin: De Gruyter, 1982); French transl., *Le Progrès scientifique* (Paris: Presses Universitaires France, 1993).

Chapter Seven

DISCOVERY DECELERATION AND THE ECONOMIC INFEASIBILITY OF PERFECTED SCIENCE

SYNOPSIS

(1) Because of the ever-escalating technological demands required for ongoing scientific progress, its advancement becomes increasingly more difficult and expensive in resource-cost terms. (2) In a world of finite resources, science must in the future progress ever more slowly—for strictly practical and ultimately economic reasons. Although natural science is theoretically limitless, its actual future development confronts obstacles and impediments of a strictly practical kind that spell its deceleration. (3) The economic demands of natural science set definite limits to what can be achieved at any state-of-the-art stage. The inherent limitations of any economically realizable state of the technological art have as a consequence the cognitive incompleteness and imperfection of its correlative science.

1. THE INFLATIONARY COSTS OF SCIENTIFIC PROGRESS

The strictly practical impediments to scientific progress that inhere in the economic aspect of its technological demands are all too often neglected by science theorists. This failure exacts penalties in terms of misunderstanding, for the fact is that significant new scientific findings are available only at the cost of expending inquiry-related resources—equipment, energy, talent, effort. In the course of progress this "purchase price" of significant new discoveries constantly increases. Actually, this state of affairs parallels that familiar arms race situation of inescapable technological obsolescence. In both cases, the economic structure is the same: technological innovation leads to exponential cost increases. (Consider the series of the B bombers, from the old B-17 of World War II, through the B-47 and B-52 of the Cold War era, to the supersonic B-1 of the 1990 era.) As science endeavors to extend its mastery over nature, it embarks on a technology-intensive arms race against nature. The escalation of technological capabilities—and, correlatively, of costs—is the manifestation of this phenomenon. We literally have a situation of cost inflation: more buys less.

This cost escalation is by no means confined to the power-intensive investigations of physics. It holds also for those areas of inquiry that—like the sciences of life or of society—are complexity intensive. Thus in biology, experiments that aim at detecting delicate statistical relationships, involving the control of large-scale animal populations, such as the "million mouse experiment" in genetics or the massive efforts to map out the DNA-coded genetic construction of biological species, indicate the lengths to which data technology will be driven in an area where massive statistics are needed. (In the social sciences, the technology of large-scale surveys affords a comparable illustration: collecting thousands of even brief interviews nowadays costs a huge sum.) The basic principles operative here are not unique to natural science. We are concerned with an endeavor to push a technology to the limits of its capacity, and one knows from innumerable cases—greater vacua, lower temperatures, higher field strengths, more powerful particle accelerations, etc.—that there is an analogous cost increase in any situation in which technology is used to press toward *any* natural limit.

All the same, the economics of science differs from that of productive enterprises of a more ordinary sort. The course of historical experience in manufacturing industries yields a picture in which (1) the industries have grown exponentially in the overall *investment* of the relevant resources of capital and labor, whereas (2) the *output* of the industries has grown at an ever-faster, indeed exponential, rate. As a result of this combination, the ratio of investment cost per unit of output has declined exponentially due to favorable "economies of scale" throughout the manufacturing industries. This relationship does *not* hold, however, for the science industry. The economics of mass production are unavailable in research at the scientific frontiers. Here, the existing modus operandi is always of limited utility: its potential is soon exhausted—the frontiers keep moving onward and upward. Of course, if it were a matter of doing a particular experiment over and over again—as is the case with classroom demonstrations—then the unit cost could be brought down and the usual economies of scale would be obtained. The economics of mass *re*production are altogether different from those in pioneering production. With mass production, the unit cost of items made exceeds that of those yet to be made.[1] Here, too, an exponential relationship obtains, but one of exponential decay. However, this is not the case with the costs of *scientific* research technology. The ratio of investment per unit of output has increased exponentially in the science industry—the exact reverse of the more standard case of industrial production.[2]

One sometimes hears the cost increase in scientific work accounted for with reference to decreasing efficiency in recruitment or to the general inflation of personal costs[3] or even to boondoggling by lazy scientists or projecteering by ambitious ones. From the present perspective, such explanatory recourse to the manpower-management side of the issue seems misguided, for the fundamental thesis to whose substantiation much of our present discussion has been dedicated is that *the primary and predominant reason for the ongoing escalation in the resource cost of significant scientific discovery resides in the increasing technical difficulties in the realization of this objective*, difficulties that are a fundamental—and ineliminable —part of the scientific enterprise itself. This economic circumstance has important consequences for the nature of scientific progress.

2. The Deceleration of Scientific Progress

In theory, the prospect of progress will never be exhausted: scientific innovation need never stop, but since ongoing progress requires us to meet ever-increasing demands, the handwriting is on the wall. We have to accept the fact that it will—and inevitably must—eventually slow down. This state of affairs is particularly clear in the context of the Law of Logarithmic Returns. If, as was argued above, scientific progress has indeed been linear in the exponential-growth past, than it will obviously have to become markedly slower in the zero-growth future.

Our deliberations indicate the reason why. Once we accept the idea that our investment in science must reach a condition of (at best) steady-state stability—and that correspondingly the accumulation of mere scientific *information* will also remain constant—then we have the situation that:

$$\frac{d}{dt} I(t) = \text{constant, and accordingly: } I(t) \approx t$$

On this basis, the volume of knowledge, which by the Law of Logarithmic Returns grows with the logarithm of the volume of information, will stand proportionate to the logarithm of elapsed time:

$$K(t) \approx \log I(t) \approx \log t$$

Consequently, the rate of increase in knowledge will stand proportionate to the *inverse* increase of the elapsed time:

$$\frac{d}{dt} K(t) \approx \frac{1}{t}$$

The constant growth of I in a steady-state era will thus be associated with a time-inversely proportional rate of growth of K. Where information grows linearly, knowledge too will grow—but at an ever-decreasing rate. Thus, the Law of Logarithmic Returns indicates that scientific progress is entering into an era of ongoing deceleration as steady-state conditions come to obtain with respect to resource investment. A zero-growth future bodes a radical and continual slowing of the pace of scientific progress.[4]

Thus if the general picture of an unremitting cost escalation in the economics of scientific progress is even partly correct, the pace of scientific advance is destined to become markedly slower in the zero-growth era that inevitably lies ahead. The half-millennium commencing around 1650 will eventually come to be regarded as among the great characteristic developmental transformations of human history, with the Age of the Science Explosion seen to be as unique—and finite—in its own historical structure as the Bronze Age and the Industrial Revolution.

3. ECONOMIC REQUIREMENTS ENTAIL LIMITATIONS

There is no escaping from the iron triangle of the real world facts that our nature-interactive technology is resource limited, that our nature-reflective data are technology limited, and that our nature-characterizing theories are data limited. So—natural science being, as it is and must be, an inescapably *empirical* enterprise—remorseless limitations are imposed upon the prospects of effective theorizing at any given stage in its development by this dependency on the available data. *Technological dependency sets technological limits* first to data acquisition and then to theory projection. To say this is not to sell human ingenuity short; it is simply a matter of facing a very fundamental fact of scientific life. Progress in the theoretical sector of natural science is insuperably limited at any given time by the implicit barriers set by the available technology of data acquisition and processing. The achieved level of sophistication in the technological state of the art of information acquisition and processing is inevitably such as to set limits to the prospects of scientific progress by restricting the observation range that is realistically accessible.

Progress without new data is, of course, possible in various fields of scholarship and inquiry. The example of pure mathematics, for instance, shows that discoveries can be made in an area of inquiry that operates without empirical data, but this hardly represents a feasible prospect for natural science. It is exactly the explicit dependency on additional data—the *empirical* aspect of the discipline—that sets natural science apart not only from the *formal* sciences (logic and mathematics) but also from the *hermeneutic* ones (such as the humanities), which address themselves ceaselessly to the imaginative reinterpretation and re-reinterpretation of old data from novel conceptual perspectives. Where natural science is concerned, progress is impracticable without new observations and new experiments.

The ancient Greeks were certainly as intelligent as we are—perhaps, arguably, even more so—but given the information-technology of the day, it is not just *improbable* but actually *inconceivable* that the Greek astronomers could have come up with an explanation for the red shift on the Greek physicians with an account of the bacteriological transmission of some communicable disease. The relevant types of data needed to put such phenomena within cognitive reach simply lay beyond their range. Given the instrumentalities of the times, *there just was no way* for the

Greeks (no matter how well endowed in brain power) to gain physical or conceptual access to the relevant phenomena. Progress in theorizing in these directions was barred, not permanently but then and there *for them*, by a technological barrier on the side of data—a barrier as absolute as the then-extant technological barriers in the way of developing the internal-combustion engine or the wireless telegraph.

The Danish historian of science, A. G. Drachmann, closes his excellent book *The Mechanical Technology of Greek and Roman Antiquity*[5] with the following observation: "I should prefer not to seek the cause of the failure of an invention in the social conditions till I was quite sure that it was to be found in the technical possibilities of the time." An important point of principle is at issue here. The history of *science*, as well as that of *technology*, is crucially conditioned by the limited nature of "the technical possibilities of the time." And these technical possibilities have unavoidable economic ramifications.

Our human acquisition of knowledge about the workings of nature is clearly a matter of *interaction*—a transaction in which *both* parties, man and nature, must play a crucial role. Most writers on the limits of science operate predominantly on the side of man and incline to see the issue as a matter of human failings and deficiencies (in intellect, learning power, memory, imagination, will power, etc.). It is too easily ignored that "limits" to scientific progress also reside in the physical limitations imposed upon us by the nature of the world.

In natural science, then, we are embarked on a potentially limitless project of improving the range of effective experimental intervention, because only by operating under new and heretofore inaccessible conditions of observational or experimental systematization—developing the power to achieve more extreme conditions of temperature, pressure, particle velocity, field strength, and so on—can we realize those circumstances that enable us to put our hypotheses and theories to the test. Our exploration of physical parameter space is, however, inevitably incomplete. We can never exhaust the whole of those parametric ranges because of increasing physical resistance as one moves toward the extremes. Thus, we inevitably face the (very real) prospect that the character of the as yet inaccessible cases generally does not conform to the patterns of regularity prevailing in the currently accessible ones. In general, new data just do not accommodate themselves to old theories. (Newtonian calculations, for example, worked marvelously for predicting solar system phenomenology—eclipses, planetary conjunctions, and the rest—but this did not mean that classical physics was free of any need for fundamental revision.) At every stage of investigative sophistication we seem to confront a different order or aspect of things. What we find in investigating nature must always in some degree reflect the character of our technology of observation, since what we are able to detect in nature is always something that depends on the mechanisms by which we search. (You do not catch little fish with a net that has a big mesh.) No matter how much we broaden that limited range of currently accessible cases, we can achieve no assurance (or even probability) that a theory corpus that smoothly accommodates the range of currently achievable outcomes will hold across the board. The prospect of future change, that is, improvement, can never be confidently foreclosed—though, to be sure, its realization becomes continually more difficult and imposes increasingly greater practical demands.

In natural science we face a situation where there is even more to be done. Further progress of a significant and substantial sort is always theoretically possible, but its realization in practice creates economic demands that cannot be met beyond a certain point. We thus encounter limits (not boundaries) to knowledge that are set not because they are hidden from our view by the nature of things but because their acquisition has cost implications that we are unwilling or unable to meet.

The demands for the enhancement of physical capacity and capability that must be met to enable further progress accordingly spell cognitive limitations for empirical science, for where there are inaccessible phenomena, there must be cognitive incompleteness as well. To this extent, at any rate, the empiricists were surely right. Only the most fanatical of rationalists could maintain the capacity of sheer intellect to compensate for lack of data. The existence of unobserved phenomena means that our theoretical systematizations may well be (and presumably are) incomplete. Insofar as various phenomena are not just undetected by us but in the very nature of the case inaccessible (even if only for *pro tem* the merely economic reasons suggested above), then our theoretical knowledge of nature must be presumed to be inadequate.[6]

As these deliberations have emphasized, scientific discovery is emphatically not cost free. In any department of science, a price must be paid for progress—and this price grows ever steeper. A point is thus bound to be reached where the cost of those further advances escalates beyond our means. In the end, scientific knowledge is something that is priceless in name only: in fact, it has a price and indeed a price that will, in the end, escalate beyond our means. The extent of our scientific knowledge is inexorably limited—not by imperfect intelligence but by the economic realities of the scientific enterprise.

Accordingly, while we can confidently anticipate that our science will undergo ongoing improvement, we cannot expect it ever to attain perfection. There is no reason to think that we ever will, or indeed can, reach the end of the road here. Further technical improvement is always in theory possible. Every successive level of technical capability has its inherent limitations, whose overcoming calls for achieving yet another, more sophisticated, level of the technological state of the art. The intensity of experimental pressures and temperatures can in principle always be increased, our accelerator-projected particles propelled closer to the speed of light, our vacua made more perfect, and so on. In all such cases, though, the inherent resistence of nature to further penetration of its parameter space imposes demands that push beyond the limits of affordability.

Given the economical and practical limitations of its development, it is clear that the imperfections of present-day science will never be completely eliminated in the future. Seeing that the scientific project is inherently incompletable—and indeed the progress toward its completion subject to logarithmic retardation

—it might be acknowledged that the science of the future will never be able to eliminate the imperfections and incompleteness of the science of the present.

If the future is anything like the past, if historical experience affords any sort of guidance in these matters, then we know that *all* of our presently favored scientific theses and theories will ultimately turn out to be untenable—that none are correct exactly as is. All the experience we can muster indicates that there is no justification for viewing our science as more than an inherently imperfect stage within an ongoing development. The ineliminable prospect of far-reaching future changes of mind in scientific matters destroys any prospect of claiming that the world is as we now claim it to be—that science's view of nature's constituents and laws is definitively correct.

In the end, then, natural science faces not barriers (insuperable boundaries of absolute limits) but obstacles (difficulties and impediments). The economic limitations that abstract the technological/economic requirements on data acquisition for scientific progress mean that we will never be able to do as much as we would, ideally speaking, like to do. Given the inescapable realities of resource limitations, the prospect of a perfected or completed science is altogether unrealistic. Science could certainly be *ended*—finished in the sense of being advanced as far as it is possible for creatures of our kind to push a venture of this sort—without thereby being *perfected or completed*, that is, without thereby discharging to the full the characterizing mandate of the enterprise in terms of description, explanation, prediction, and control.

Thus, the technological—and thus ultimately *economic*—ramifications of science as a human project also have major implications for science as a cognitive discipline. This important consideration warrants closer scrutiny.[7]

NOTES

1. Compare Keith Norris and John Vaizey, *The Economics of Research and Technology* (London: Allen & Unwin, 1973), p. 164.

2. This latter relationship has been verified for many industries, including the production of fuels, chemicals, and metals. See, for example, the data cited in M. Korach, "The Science of Industry," in *The Science of Science*, ed. M. Goldsmith and A. Mackey (London, 1964), pp. 179-94.

3. Derek J. deSolla Price, *Little Science, Big Science* (New York: Columbia University Press, 1963), pp. 92-93 and 101ff.

4. This relationship conveys an important lesson, for the question arises: Is the situation of diminishing returns on scientific effort not incompatible with the fact (so decidedly emphasized in Chapter Two) that natural science is potentially incompletable? To see that no incompatibility arises, it suffices to recall that an ever-decreasing series need not yield a *convergent* sum—as indeed the series 1/2, 1/3, 1/4, does not, and it is just this series that corresponds to the relationship:

$$\frac{d}{dt} K(t) = \frac{l}{t}$$

5. Copenhagen and Madison, 1963.

6. This chapter's themes are also addressed in the author's *Scientific Progress* (Oxford: Basil Blackwell, 1978). The book is also available in translation: German transl., *Wissenschaftlicher Fortschitt* (Berlin: De Gruyter, 1982); French transl., *Le Progrès scientifique* (Paris: Presses Universitaries de France, 1994).

7. On this chapter's theme see also the author's *The Limits of Science* (Berkeley and Los Angeles: University of California Press, 1984). Some aspects of the issue are also interestingly discussed in Roger Trigg, *Rationality and Science: Can Science Explain Everything?* (Oxford: Blackwell, 1993).

Chapter Eight

HUMAN SCIENCE AS CHARACTERISTICALLY HUMAN

SYNOPSIS

(1) In theory, very different versions of "natural science" are possible in a universe where very different sorts of extraterrestrial beings might possibly exist. (2) The reasoning that there must be one single science because there is one single world is gravely defective. (3) A quantitative analysis of the situation indicates that our particular human version of natural science may well be something unique to our particular situation. (4) Moreover, it is quite wrong to think that potentially different versions of science are all positioned along one single developmental route. (5) Natural science as we know it may well be—indeed presumably is—a characteristically human enterprise, specifically reflecting our particular manner of using the material and intellectual resources at our disposal.

1. THE POTENTIAL DIVERSITY OF "SCIENCE"

To what extent does the investment of specifically human effort and energy condition the character of our natural science? Does our science as a product reflect our particular modus operandi? The present chapter will consider this issue through the perspective of the question of whether an astronomically remote civilization might be scientifically more advanced than ourselves. On this basis, our fundamental thesis that science should be viewed in economic perspective as a productive enterprise of our culture can be clarified in a vivid and illuminating way.

The seemingly straightforward question about the possibility of scientifically more advanced aliens turns out on closer inspection to involve considerable complexity. This complexity relates not only to the actual or possible facts of the situation, but also—and crucially—to theoretical questions about the very ideas or concepts that are at issue here. To begin with, there is the problem of just what it is for there to be another science-possessing civilization. Note that this is a question that *we* are putting—a question posed in terms of the applicability of *our* term "science." It pivots on the issue of whether *we* would be prepared to call certain of *their* activities—once we came to understand them—as engaging in scientific inquiry, and whether we would be prepared to recognize the product of these activities as constituting a (state of a branch of) science.

A scientific civilization is not merely one that possess intelligence and social organization, but one that puts these resources to work in a certain very particular sort of way. This opens up the rather subtle issue of priority in regard to process vs. product. We must resolve whether what counts for a civilization's "having a science" is primarily a matter of the substantive *content* of their doctrines (their belief structures and theory complexes) or is primarily a matter of the aims and *purposes* with which their doctrines are formed.

The matter of content turns on the issue of how similar their scientific beliefs are to ours. A look at our own historical evolution indicates that this is clearly something on which we would be ill advised to put much emphasis at the very outset. After all, the speculations of the nature theorists of pre-Socratic Greece, our ultimate ancestor in the scientific enterprise, bear precious little resemblance to our present-day sciences, nor does contemporary physics bear all that much doctrinal resemblance to that of Newton. It is clearly more appropriate to give prime emphasis to matters of process and purpose.

Accordingly, the matter of these aliens "having a science" is to be regarded as turning not on the extent to which their substantive *findings* resemble ours, but on the extent to which their purposive *project* resembles ours—of deciding whether we are engaged in the same sort of inquiry in terms of the sorts of issues being addressed and the ways in which they are going about addressing them. The issue accordingly is at bottom not one of the substantive similarity of their scientifically formed beliefs to ours, but one of the functional equivalency of the *projects* at issue in terms of the quintessential goals that define the scientific enterprise as what it is: explanation, prediction, and control over nature. It is this that ultimately defines what it is for those aliens to have a science.

This perspective enjoins the pivotal question: To what extent would the *functional equivalent* of natural science built up by the inquiring intelligences of an astronomically remote civilization be bound to resemble our science in substantive content-oriented regards? In considering this issue, one soon comes to realize that there is an enormous potential for diversity here.

To begin with, the *machinery of formulation* used by an alien civilization in expressing its science might be altogether different. In particular, the aliens' mathematics might be very unlike ours. Their "arithmetic" could be anumerical—purely comparative, for example, rather than quantitative. Especially if their environment were not amply endowed with solid objects or stable structures—if, for example, they were jellyfishlike creatures swimming about in a soupy sea—their "geometry" could be something rather strange, largely topological, say, and geared to structures rather than sizes or shapes. Digital thinking in all its forms might be undeveloped, while they might, like the ancient Greeks, have "Euclidean" geometry without analysis. Thus, seeing that the mathematical mechanisms at their disposal could be very different from ours, it is clear that their description of nature in mathematical terms could also be very different (not necessarily truer or falser, but just different.)

Second, the *orientation* of the science of an alien civilization might be very different. All the aliens' efforts might conceivably be devoted to the social sciences—to developing highly sophisticated analyses of psychology and sociology, for example. Again, their approach to natural science might also be very different. Communicating by some sort of "telepathy" based upon variable odors or otherwise "exotic" signals, they might devise a complex theory of thought-wave transmittal through an ideaferous aether. Electromagnetic phenomena might lie altogether outside their ken; if their environment does not afford them loadstones and electrical storms, etc., the occasion to develop electromagnetic theory might never arise. The course of scientific development tends to flow in the channel of practical interests. A society of porpoises might lack crystallography but develop a very sophisticated hydrodynamics; one comprised of molelike creatures might never dream of developing optics. The science of a different civilization would presumably be closely geared to the particular pattern of their interaction with nature as funneled through the particular course of their evolutionary adjustment to their specific environment. Alien civilizations might scan nature very differently. The direct chemical analysis of environmental materials might prove highly useful to them, with bioanalytic techniques akin to our sense of taste and smell highly developed so as to provide the basis for a science of a very different sort. Acoustics might mean very little to them, while other sorts of pressure phenomena—say the theory of turbulence in gases—might be the subject of intense and exhaustive investigation. Rather than sending signals by radio waves or heat radiation signals, they might propel gravity waves through space. After all, a comparison of the "science" of different civilizations here on earth suggests that it is not an outlandish hypothesis to suppose that the very *topics* of an alien science might differ radically from those of ours. In our own case, for example, the fact that we live on the surface of our planet (unlike whales or porpoises), the fact we have eyes (unlike worms or moles) and thus can *see* the heavens, the fact that we are so situated that the seasonal positions of heavenly bodies are intricately connected with our biological needs through the agricultural route to food supply are all clearly connected with the development of astronomy. Accordingly, the

constitution of the alien inquirers—physical, biological, and social—emerges as a crucial element here. It
serves to determine the agenda of questions and the
instrumentalities for their resolution—to fix what
counts as interesting, important, relevant, significant.
In determining what is seen as an appropriate question and what is judged as an admissible solution, the
cognitive posture of the inquirers must be expected to
play a crucial role in shaping and determining the
course of scientific inquiry itself.

Third, the *conceptualization* of an alien science
might be very different. We must reckon with the
theoretical possibility that a remote civilization might
operate with a radically different system of concepts
in its cognitive dealings with nature. To motivate this
idea of a conceptually different science, it helps to cast
the issue in temporal rather than spatial terms. The
descriptive characterization of *alien* science is a project rather akin in its difficulty to that of describing
our own *future* science. As we saw in Chapter Five, it
is thus effectively impossible to predict not only the
answers but even the questions that are on the agenda
of future science, because these questions will grow
out of the answers we obtain at yet unattained stages
of the game. The situation of an alien science could be
much the same. As with the science of the remote
future, the science of the remotely distant must be
presumed to be of such a nature that we really could
not achieve intellectual access to it on the basis of our
own position in the cognitive scheme of things. Just
as the technology of another highly advanced civilization would most likely strike us as magic, so its science
would most likely strike us as incomprehensible gib-

berish—until we had learned it from the ground up. They might (just barely) be able to *teach* it to us, but they almost certainly could not *explain* it to us. After all, the most characteristic and significant sort of difference between variant conceptual schemes arises when the one scheme is committed to something the other does not envisage at all—something that lies outside the conceptual range of the other. The science of different civilizations, may well, like Galenic and Pasteurian medicine, in key respects simply *change the subject* so as no longer to talk about the same things, but treat things (e.g., humors and bacteria, respectively) of which the other takes little or no cognizance at all. If, for example, certain intelligent aliens should prove a diffuse and complex aggregate mass of units comprising wholes in ways that allow overlap (compare Ehrensvärd, 1965, pp. 146-148–see the Select Bibliography on pp. 154ff for reference of this format), then the role of social concepts might become so paramount that nature as a whole comes to be viewed in these terms. The result would be something very difficult for us to grasp, seeing that they are based on a mode of shared experience with which we have no contact.

It is only reasonable to presume that the conceptual character of the (functionally understood) science of an alien civilization is so radically different from ours in substantive regards as to orient the aliens' thought about the nature of things in altogether different directions. Their approach to classification and structurization, their explanatory mechanisms, their predictive concerns, and their modes of control over nature might all depart significantly from our own.

Natural science—broadly construed as inquiry into the ways of nature—is something that is in principle almost infinitely plastic. Its development will trace out an historical course that is bound to be closely geared to the specific capacities, interests, environment, and opportunities of the creatures that develop it. We are deeply mistaken if we think of it as a process that must follow a route roughly parallel to ours and issue in a comparable product. It would be grossly unimaginative to think that either the journey or the destination must be the same—or even substantially similar.

2. THE ONE WORLD, ONE SCIENCE ARGUMENT

One recent writer on extraterrestrials asks the question, "What can we talk about with our remote friends?" and answers it with the remark: "We have a lot in common. We have mathematics in common, and physics, and astronomy." (E. Purcell in Cameron, 1963, p. 142.) This line of thought begs some very big questions.

Our alien colleagues scan nature for regularities using (at any rate to begin with) the sensors provided them by their evolutionary heritage. They note, record, and transmit those regularities that they find to be intellectually interesting or pragmatically useful. They develop their inquiries by theoretical triangulation that proceeds from the lines indicated by these resources. Now this is clearly going to make for a course of development that closely gears their science to their particular situation—their biological endowment, (their sensors), their cultural heritage (what is interesting), their environmental niche (what is pragmatically useful). Where these key parameters differ, we can confidently expect that the course of scientific development will differ as well.

Admittedly, there is only one universe, and its laws, as best we can tell, are everywhere the same. So if intelligent aliens investigate nature at all, they will investigate the same nature we ourselves do, but the sameness of the object of contemplation does nothing to guarantee the sameness of the ideas about it. It is all too familiar a fact that even where human (and thus *homogeneous*) observers are at issue, different constructions are often placed upon "the same" occur-rences. Primitive peoples thought the sun a god and the most sophisticated among the ancient peoples thought it a large mass of fire. We think of it as a large thermonuclear reactor, and heaven only knows how our successors will think of it in the year 3000. As the course of human history clearly shows, there need be little uniformity in the conceptions held about one self-same object by differently situated groups of thinkers.

It is surely naive to think that because one selfsame object is in question, its description must issue in one selfsame result. This view ignores the crucial matter of one's intellectual orientation. One selfsame piece of driftwood is viewed very differently indeed by the bota-nist, the painter, the interior decorator, the chemist, the woodcarver, etc. It's a matter of what aspect of the thing is focused upon as important or interesting. Minds with different sorts of concerns and interests and different backgrounds of information can deal with mutually common items in ways that yield wholly disjoint and disparate results because altogether dif-ferent features of the thing are being addressed. It is notorious that observers are prisoners of their cognitive preparation, interests, and predispositions—seeing only what their preestablished cognitive resources enable them to see and blind to that for which they are cognitively unprepared.

With science, as with any productive enterprise, it is not only the raw material but also the mode of productive processing that serves to determine the nature of the outcome.

Accordingly, the sameness of nature and its laws by no means settles the issue of scientific uniformity, for science is always the result of *inquiry* into these matters and this is inevitably a matter of a *transaction* or *interaction* in which nature is but one party and the inquiring beings another. The result of such an interaction depends crucially on the contribution from both sides—from nature and from the intelligences that interact with it. A kind of chemistry is at issue, where nature provides only one input and the inquirers themselves provide another—one that can massively and dramatically affect the outcome in such a way that we cannot disentangle the respective contributions of the two parties, nature and the inquirer.

Each inquiring civilization must be thought of as producing its own, itself ever-changing cognitive product—all more or less adequate in their own ways—but with little if any actual overlap in conceptual content. Human organisms are essentially similar, but there is not much similarity between the medicine of the ancient Hindus and that of the ancient Greeks. There is every reason to think that the natural science of different astronomically remote civilizations should be highly diversified. Even as different creatures can have a vast variety of lifestyles for adjustment within one selfsame physical environment such as this earth, so can they have a vast variety of thought styles for cognitive adjustment within one selfsame world.

After all, throughout the earlier stages of man's intellectual history, different human civilizations have developed their natural sciences in a substantially different way. The shift to an extraterrestrial perspective is bound to amplify such cultural differences. Perhaps reluctantly, we must face the fact that on a cosmic scale the hard physical sciences have something of the same cultural relativity that one encounters with the material of the softer social sciences on a terrestrial basis.

It seems reasonable to argue: "Common problems constrain common solutions. Intelligent alien civilizations have in common with us the problem of cognitive accommodation to a shared world. Natural science as we know it is our solution of this problem. Ergo, it is likely to be theirs as well." This tempting argument founders, however, on its second premise. Their problem is *not* common with ours because their situation must be presumed substantially different, seeing that they live in a significantly different environment and come equipped with significantly different resources. To presuppose a common problem is in fact to beg the question.

There is no quarrel here with the principle of the uniformity of nature, but this principle merely tells us that when exactly the same question is put to nature, exactly the same answer will be forthcoming. The development of a science, however, hinges crucially on this matter of questions—to the sorts of issues that are addressed and the sequential order in which they are posed. Here the prospect of variation arises: We must expect alien beings to question nature in ways very different from our own. On the basis of an *interactionist* model, there is no reason to think that the sciences of different civilizations will exhibit anything more than the roughest sorts of family resemblance.

Our science reflects not only our interests but also our capacities. It addresses a range of issues that are correlative without specific modes of physical interaction with nature, the specific ways in which we monitor its processes. It is highly selective—the science of a being that gets its most crucial information through sight, monitoring developments along the spectrum of electromagnetic radiation, rather than, say, monitoring variations of pressure or temperature. Ours is certainly not a phenomenalistic science geared to the feel of things or the taste of things. The science we have developed depends on our capacities and needs, our evolutionary heritage as beings inserted into the orbit or natural phenomena in a certain particular way.

The fact is that all such factors as capacities, requirement, interests, and course of development affect the shape and substance of the science and technology of any particular place and time. Unless we narrow our intellectual horizons in a parochially anthropomorphic way, we must be prepared to recognize the great likelihood that the science and technology of another civilization will be something *very* different from science and technology as we know it. We are led to view that our human sort of natural science may well be sui generis, adjusted to and coordinate with a being of our physical constitution, inserted into the orbit of the world's processes and history in our sort of way. It seems that in science as in other areas of human endeavor we are prisoners of the thought world that our biological and social and intellectual heritage affords us.

3. A QUANTITATIVE PERSPECTIVE

Let us attempt to give some quantitative structure to the preceding qualitative deliberations by bringing some rough order-of-magnitude estimates on the scene.

First off, there is the problem of estimating H, the number of habitable planets in the universe. This assessment can be formed by means of the following quantities, themselves represented merely as order-of-magnitude specifications:

n_1 = number of galaxies in the observable universe (10^{11}).

n_2 = average number of star systems per galaxy (10^{11}).

x_1 = fraction of star systems having suitably large and stable planets $(1/10)$.

n_3 = average number of such planets in the temperature zone of a suitably benign solar system, where it is neither too hot nor too cold for life (1).

x_2 = fraction of temperature planets equipped with a surface chemistry capable of supporting life $(1/10)$.

These figures—borrowed in the main from Dole, 1970, and Sagan, 1980—can be subject to scepticism. They, and those that are to follow, must be viewed realistically. They are not graven in stone for all the ages, but represent conjectural "best estimates" in the present state of the art. The important point, as will emerge below, is that the overall tendency of our discussion is not acutely sensitive to precision in this respect. Accordingly, one should look on the calculations that are to follow as rather suggestive than in any sense conclusive. Their function is to indicate a general line of thought, and not to establish a definitive conclusion.

Given the preceding estimates, the sought-for number of habitable planets will be the product of these quantities:

$$H = 10^{20}$$

This, of course, is a prodigiously large number, providing for some thousand million habitable planets per galaxy (cf. Dole, 1970, p. 103). We here confront a truly impressive magnitude, but it is only the start of the story.

A planet capable of supporting life might well have no life to support, let alone *intelligent* life. The point is that the physics, chemistry, and biology must all work out just right. The physical, chemical, and biological environments must all be duly auspicious and exactly the right course of triggering processes must unfold for the evolution of intelligence to run a successful course. Our next task is thus to estimate I, the number of planets on which intelligent life evolves. Let us proceed here via the following (again admittedly rough and ready) quantities:

r_1 = fraction of habitable planets on which life—that is, some sort of self-reproducing biological system—actually arises (1/100).

r_2 = fraction of these on which highly complex life-forms evolve, possessed of something akin to a central nervous system and thus capable of complex (though yet instinctively programmed) behavior forms (1/100).

r_3 = fraction of these on which intelligent and sociable beings evolve—beings who can acquire, process, and exchange factual information with relative sophistication—who can observe, remember, reason, and communicate (1/100).

As these fractions indicate, the evolutionary process that begins with the inauguration of life and moves on to the development of intelligence is certainly not an inexorable sequence, but one that could, given suitably inauspicious conditions, abort in a stabilization that freezes the whole course of development at some plateau along the way. Note that r_3 in particular involves problems. Conscious and indeed even intelligent creatures are readily conceivable who yet lack that orientation toward their environment needed to acquire, store, transmit, and process the factual information necessary to science. Where such conceptions as space, time, process, unit, function, and order are missing, it is difficult to see how anything deserving of the name "science" could exist. An intelligence unswervingly directed at the aesthetic appreciation of particular phenomena rather than their generally lawful structure is going to miss out on the scientific dimension.

So, when we put the fractions of the preceding series to work, we arrive at

$$I = 10^{14}$$

This unquestionably still indicates an impressively large number of intelligence-bearing planets. It would, in fact, yield a quota of some thousand per galaxy (a figure that, if, correct, would cast a shadow over the prospect of our ever establishing contact with extraterrestrial intelligence, since it would indicate its nearest locale to be some 1,000 light-years away).

As regards this figure, one can say that it would certainly be possible to take a more rosy view of the matter. One could suppose that nature has a penchant for life—that a kind of Bergsonian *élan vital* is operative, so that life springs forth wherever it can possibly get a foothold. Something of this attitude certainly underlies J. P. T. Pearman's contention (Cameron, 1963, p. 290) that the probability is 1 that life will develop on a planet with a suitable environment—a stance in which MacGowan and Ordway, 1966 (p. 365), Dole, 1970 (pp. 99-100), and Ball, 1973 (p. 347) concur. (Sagan, 1980, p. 300, is slightly more conservative in fixing this quantity at 1/3.) One theorist cuts the Gordian Knot with a curious bit of reasoning:

> Biological evolution proceeds by the purely random process of mutation . . . . Since the process is a random one, the laws of probability suggest that the time-scale of evolution on earth should resemble the average time-scale for the development of higher forms of life anywhere. (Huang, 1960, p. 55.)

This blatantly ignores the crucially differentiating role of initial conditions in determining the outcome of random processes. The terrain through which a random walk proceeds is going to make a lot of difference to its destination. The transition from habitability to habitation—from the possibility of life to its actuality—is surely not all that simple. Sir Arthur Eddington did well to remind us in this context of the prodigality of nature when he asked how many acorns are scattered for any one that grows into an oak. (Eddington, 1928, p. 177).

One could perhaps go on to suppose that nature incorporated a predisposition for intelligence—that there is a Teilhard-de-Chardin-reminiscent impetus towards *nous*, so that intelligence develops wherever there is life. Indeed the suggestion is sometimes made in this vein that "the adaptive value of intelligence . . . is so great . . . that if it is genetically feasible, natural selection seems likely to bring it forth." (Shklovskii and Sagan, 1966, p. 411.) This argument from utility to evolutionary probability clearly has its limitation. ("[T]here are no organisms on Earth which have developed tractor treads for locomotion, despite the usefulness of tractor treads in some environments." *Ibid.*, p. 359.) Moreover, this suggestion seems implausibly anthropocentric. To all appearances the termite has a securer foothold on the evolutionary ladder than man; and the coelacanth can afford to smile when the survival advantages of intelligence are touted by a johnny-come-lately creature whose self-inflicted threats to long-term survival are a cause of general concern. As J. P. T. Pearman has rightly noted, "the successful persistence of a multitude of simpler organisms from ancient times argues that intelligence may confer no unique benefits for survival in an environment similar to that of earth" (Cameron, 1963, p. 290). After all, it will prove survival-conductive mainly for a being of a particularly restless disposition, a creature such as man, who refuses to settle down in a secured ecological niche, but shifts restlessly from environment to environment needing continually to readjust to self-imposed changes. The value of intelligence, one might say, is not absolute but remedial—as an aid to offsetting the problems of a particular sort of lifestyle. We would do well to think of the emergence of intelligence as a long series of fortuitous twists and turnings rather than an inexorable push toward a foreordained result. It would be glib in the extreme to assume that once life arises, its subsequent development would proceed in much the same way as here on earth. (Simpson, 1964, provides a useful perspective here.)

The indicated figures accordingly seem plausibly middle-roadish between undue pessimism and an intelligence-favoring optimism that seems unwarranted at this particular stage of the scientific game. Even so, it is clear that the proposed specification of I represents a strikingly substantial magnitude—one that contemplates many thousand of millions of planets equipped with intelligent creatures scattered throughout the universe.

Intelligence, however, is not yet the end of the line. (After all, dolphins and apes are presumably intelligent, but they do not have, and are unlikely to develop, a science.) Many further steps are needed to estimate S, the number of planets throughout the universe in which scientific civilizations arise. The developmental path from intelligence to science is a road strewn with substantial obstacles. Here matters must be propitious not just as regards the physics and chemistry and biochemistry and evolutionary biology and cognitive psychology of the situation. The social-science requisites for the evolution of science as the cultural artifact of a multifocal civilization must also be met. Economic conditions, social organization, and cultural orientation must all be properly adjusted before the move from intelligence to science can be accomplished. For scientific inquiry to evolve and flourish there must, in the first place, be cultural institutions whose development requires economic conditions and a favorable social organization. Terrestrial experience suggests that such conditions for the social evolution of a developed culture are by no means always present where intelligence is. We do well to recall that of the myriad human civilizations evolved here on earth only one, the Mediterranean/European, managed to develop natural science in a form unproblematically recognizable as such. The successful transit from intelligence to science is certainly not a sure thing.

Let us once more look at the matter quantitatively:

p_1 = probability that intelligent beings will (unlike dolphins) also possess developed manipulative abilities and will (unlike the higher apes) combine intelligence with manipulative ability so as to develop a technology that can be passed on as a social heritage across the generations. (.01)

p_2 = probability that technologically competent intelligent beings will group themselves in organized societies of substantial complexity—a transition that stone-age man, for example, never managed to make. (.1)

p_3 = probability that an organized society will not only acquire the means for transmitting across successive generation the political and pragmatic know-how indispensable to an "organized society" as such, but will also (unlike the ancient Egyptians) develop institutions of learning and culture for accumulating, refining, systematizing, and perpetuating factual information. (.1)

p_4 = probability that society with cultural institutions will develop an unstable (i.e., continually developing and dynamic) technology—in a way the ancient Greeks and the old Chinese mandarins, for example, never did—so as to create a technologically progressive civilization. (.01)

p_5 = probability that a technologically progressive civilization will develop and maintain an articulated "science" and concern itself with the theoretical study of nature at a level of high generality and precision. (.1)

However firm the physical quantities are with which we began, we are by now skating on very thin ice indeed. These latter issues of sociology and cognitive psychology can be quantified only in the most tentative and cautious way. The one thing that is clear is that a good many conditions of this sort have to be met and that each involves a likelihood of relatively modest proportions.

The issue of technology reflected in p_1 and p_4 is particularly critical here. The urge to an ever-aggressive technological extension of self is certainly not felt by every intelligent life form. It is a part of Western man's peculiar lifestyle to impatiently cultivate the active modification of nature in the pursuit of human convenience so as to create an artificial environment of ultra-low entropy. Even in human terms this is not a uniquely constrained solution to the problem of evolutionary adaptation. Many human societies seem to have remained perfectly content with the status quo for countless generations and very sophisticated cultural projects—literary criticism, for example—have developed in directions very different from the scientific. After all, a culture can easily settle comfortably into a frozen traditional pattern with respect to technology. (If their attention span is long enough, our aliens might cultivate scholastic theology *ad indefinitum.*) Moreover, unless their oral lore is something very different from ours, it is hard to see how an alien civilization could develop science without writing—a skill that even many human communities did not manage to develop. The salient point is that for science to emerge on a distant planet it is not enough for there to be life and intelligence; there must also be culture and progressive technology and explanatory interest and theorizing competency.

The product of the preceding sequence of probability estimates is 10^{-7}. Multiplying this by I we would obtain the following expected-value estimate of the number of science-possessing planetary civilizations:

$$S = 10^7 = 10,000,000$$

This, of course, is still a large number, albeit now one that is rather modest on a cosmic scale, implying a chance of only some .01 percent that a given galaxy actually provides the home for a science. (Note too that we ignore the temporal dimension—the scientific civilizations at issue may have been destroyed long ago, or perhaps simply have lost interest in doing science.) A more conservative appraisal of the sociological parameters has thus led us to a figure that is more modest by many orders of magnitude than the estimate by Shklovskii and Sagan (1966, p. 418) that some $10^{5\pm1}$ scientifically sophisticated civilizations exist in our galaxy alone. Nevertheless, even our modest ten million is a sizable number.

4. COMPARABILITY AND JUDGMENTS OF RELATIVE ADVANCEMENT OR BACKWARDNESS

Let us now come to grips with the crux of our present concerns—the issue of scientific advancement. Earlier we defined "science" in terms of a rather generic sort of *functional* equivalency. The question, however, from which we began was not whether a remote civilization has a science of some sort, but whether it is *scientifically more advanced* than ourselves. Now *advancement* is to be at issue, the question is not one of the relative volume of intelligence as such—with data processing in volumetric terms—but one of the quality of the orientation of intelligence towards substantive issues. If another *science* is to represent an advance over ours, we must clearly construe it as *our sort* of science in rather particularized and substantive terms. But given the immense diversity to be expected among the various modes of science and technology, the number of extraterrestrial civilizations possessing a science and technology that are duly consonant and contiguous with ours—and in particular, heavily geared toward the mathematical laws of the electromagnetic spectrum—must be judged to be very small indeed.

We have come to recognize that sciences can vary (1) in their formal mechanisms of *formulation*—their mathematics, (2) in their *conceptualization*, that is, in the kinds of explanatory and descriptive concepts they bring to bear, and (3) in their *orientation* toward the manifold pressures of nature, reflecting the varying interest directions of their developers. While "science" as such is clearly not anthropocentric, science *as we have it*—the only "science" that we ourselves know—is a specifically human artifact that must be expected to reflect in significant degree the particular characteristics of its makers. Consequently, the prospect that an alien science possessing civilization has a *science* that we would acknowledge (if sufficiently informed) as representing the same general line of inquiry as that in which we ourselves are engaged seems extremely implausible. The possibility that *their* science and technology are sufficiently similar in orientation and character to substantively approximate *ours* must be viewed as extremely remote. We clearly cannot estimate this as representing something other than a very long shot indeed—certainly no better than one in many thousands.

Just such comparability with our sort of science is, however, the indispensable precondition for judgments of relative advancement or backwardness vis-à-vis ourselves. The idea of their being scientifically more advanced is predicated on the uniformity of the enterprises—doing better and more effectively the kinds of things that *we* want science and technology to do. Any talk of advancement and progress is predicated on the sameness of the direction of movement: only if others are traveling along the same route as we can they be said to be ahead of or behind us. The issue of relative advancement is linked inseparably to the idea of doing the same sort of thing better or more fully. This falls apart when "this sort of thing" is not substantially the same. One can say that a child's expository writing is more primitive than an adult's or that the novice's performance at arithmetic or piano playing is less developed than that of the expert, but we can scarcely say that Chinese cookery is more or less advanced than Roman, or Greek pottery than Renaissance glass-blowing. The salient point for present purposes is simply that where the enterprises are sufficiently diverse, the ideas of comparative advancement and progress are inapplicable for lack of a sine qua non condition.

Claiming scientific superiority is not as simple as may seem at first sight. To begin with, it would not automatically emerge from the aliens' capacity to make many splendidly successful predictions, for this could be the result of precognition or empathetic attunement to nature or such-like. Again, what is wanted is not just a matter of *correct*, but of cognitively underwritten and thus *science-guided* predictions— predictions guided by insight based on understanding and not mere lucky guesswork—and that's just exactly what is to be proved.

It clearly is not enough for establishing their being scientifically more advanced than ourselves that the aliens should perform technological wonders—that they should be able to do all sorts of things we would like to do but cannot. After all, bees can do that. The technology at issue must clearly be the product of intelligent contrivance rather than evolutionary trial and error. What is needed for advancement is that their performatory wonders issue from superior theoretical knowledge—that is, from superior science. And then we are back in the circle.

Nor would the matter be settled by the consideration that an extraterrestrial species might be more intelligent than we are in having a greater capacity for the timely and comprehensive monitoring and processing of information. After all, whales or porpoises, with their larger brains, may (for all we know) have to manipulate relatively larger quantities of sheer data than ourselves to maintain effective adaptation within their highly changeable environment. What clearly counts for scientific knowledge is not the *quantity* of intelligence in sheer volumetric terms but its *quality* in substantive, issue-oriented terms. Information handling does not ensure scientific development. Libraries of information (or misinformation) can be generated about trivia—or dedicated to matters very different from science as we know it.

It is perhaps too tempting for humans to reckon cognitive superiority by the law of the jungle—judging as superior those who do or would come out on top in outright conflict. Surely the Mongols were not possessors of a civilization superior to that of the Near Eastern cultures they overran. Again, we earthlings might easily be eliminated by not very knowledgeable creatures able to produce at will—perhaps by using natural secretions—a biological or chemical agent capable of killing us off.

The key point, then, is that if they are to effect an *advance* on our science, they must both (1) be engaged in doing roughly our sort of thing in roughly our sort of way and (2) do it significantly better. In speaking of the science of another civilization as "more advanced" than our own, we contemplate the prospect that they have developed *science* (*our* sort of science— science as we know it) further than we have ourselves. This is implausible. Even assuming that they develop a science at all—that is, a *functional equivalent* of our science—it seems unduly parochial to suppose that they are at work constructing *our* sort of science in substantive, content oriented terms. Diverse life modes have diverse interests; diverse interests engender diverse technologies; diverse technologies make for diverse modes of science. Where the parties concerned are going in different directions, it makes no sense to say that one is ahead of or behind the other.

If a civilization of intelligent aliens develops a science at all, it seems plausible to expect that they will develop it in another direction altogether and produce something that we, if we could come to understand it at all, would regard as simply detached in a content orientation—though perhaps not in intent—from the scientific enterprise as we ourselves cultivate it. (Think of the attitude of orthodox sciences to exotic phenomena such as hypnotism or acupuncture, let alone to parapsychology.)

The crucial consideration is that there just is no single-track itinerary of scientific/technological development that different civilizations travel in common with mere differences in speed or in staying power (notwithstanding the penchant of astrophysicists for the neat plotting of numerical degrees of development against time in the evolution of planetary civilizations—cf. Ball, 1980, p. 658). In cognition and even in scientific evolution we are not dealing with a single-track railway, but with a complex network leading to many mutually remote destinations. Even as cosmic evolution involves a red shift that carries different star systems ever farther from each other in space, so cognitive evolution may well involve a red shift that carries different civilizations even farther from each other into mutually remote thought worlds.

The prospect that an alien civilization is going about the job of doing *our* science—a science that reflects the sorts of interests and involvement that *we* have in nature—better than we do ourselves must accordingly be adjudged as extremely far-fetched. Specifically, two conditions would have to be met for the science of an intelligent civilization to be in a position to count as comparable to ours:

(1) That, *given* that they have a science and a developing technology, they have managed to couple the two and have proceeded to develop (unlike the ancient Greeks and Chinese and Byzantines) a *science-guided* technology. (Probability p_6.)

(2) That their science-guided technology is sufficiently oriented toward issues regarding natural processes sufficiently close to those at which our science-guided technology is oriented that a comparison can reasonably be made between them. (Probability p_7.)

To judge by terrestrial experience, it seems rather optimistic to estimate p_6 to be even so large as one in a thousand (with $p_6 = .001$); and p_7 must also be adjudged as quite small. As we have seen, science-guided technology could be oriented in very different directions. The potential diversity of different modes of science is encrimous, so that there is little choice but to see p_7 as an eventuation whose chances are no better than, say, one in ten thousand (so p_7 = .0001). If our alien scientists are differently constructed (if they are silicon-based creatures, for example) or if their natural environment is very different, their practical interests and accordant technology will be oriented in very different directions from ours. For example, their technology might be wholly independent of hardware, oriented not toward physical machinery, but toward the software of mind-state manipulation, telepathy, hypnotism, autosuggestion, or the like. (Ray Bradbury's Martians destroy an expedition from earth armed with atomic weapons by thought control.) We must not keep our imagination on a short leash in this regard. Given the diversity of different modes of science and the enormous spectrum of possible issues and purposes in principle available to extraterrestrial aliens, the prospect must be recognized that the direction of their science-guided technology might be vastly different from ours.

Accordingly we have it that

$$p_6 \times p_7 = 10^{-7}$$

Now, the product of this quantity with the quantity S, the number of civilizations that possess a technologized science as we comprehend it, is clearly not going to be very substantial—it is, in fact, going to be strikingly close to 1.

If "being there" in scientific regards means having *our* sort of scientifically guided technology and our sort of technologically channeled science, then it does not seem all that far-fetched to suppose that as regards science as we have it *we might be there alone*—even in a universe amply furnished with other intelligent civilizations. The prospect that somebody else could do our sort of thing in the scientific sphere better than we can do it ourselves seems very remote.

5. BASIC PRINCIPLES

The overall structure of our analysis thus emerges in Table 1. Its figures interestingly embody the familiar situation that as one moves along a nested hierarchy of increasing complexity, one encounters a greater scope of diversity—that the further layers of system complexity provide for an ever-widening spectrum of possible states and conditions. (The more fundamental that system, and narrow its correlative range of alternatives, the more complex and the wider.) If each unit (letter, cell, atom) can be configurated in ten way, then each ordered group of ten such units (word, organ, molecule) can be configurated in 10^{10}, and each complex of ten such groups (sentences, organisms, objects) in $(10^{10})^{10} = 10^{100}$ ways. Thus, even if only a small fraction of what is realizable in theory is realizable in nature, any increase in organizational complexity will nevertheless be accompanied by an enormous amplification of possibilities.

Table 1

CONDITIONS FOR THE DEVELOPMENT OF SCIENCE

planets of sufficient size for potential habitation	10^{22}
fraction thereof with affording:	
temperate location for	10^{-1}
chemistry for life support	10^{-1}
biochemistry for the actual emergence of life	10^{-2}
biology and psychology for the evolution of intelligence	10^{-4}
sociology for developing a culture with a technology and a science	10^{-7}
epistemology for developing science as we know it	10^{-7}

To be sure, the numerical particulars that constitute the quantitative thread of the discussion cannot be given much credence. Their general tendency nevertheless conveys an important lesson, for people frequently seem inclined to reason as follows:

> There are after all, an immense number of planetary objects running about in the heavens. And proper humility requires us to recognize that there is nothing at all that special about the Earth. If it can evolve life and intelligence and civilization and science, then so can other planets. And given that there are so many other runners in the race we must assume that—even though we cannot see them in the cosmic darkness— some of them have got ahead of us in the race.

As one recent writer formulates this familiar argumentation, "Since man's existence on the earth occupies but an instant in cosmic time, surely intelligent life has progressed far beyond our level on some of these 100,000,000 (habitable) planets (in our galaxy)" (M. Calvin in Cameron, 1963, p. 75), but such plausible-sounding argumentation overlooks the numerical complexities. Even though there are an immense number of solar systems, and thus a staggering number of planets (some 10^{22} on our estimate), nevertheless, a substantial number of conditions must be met for science (as we understand it) to arise. The astrophysical, physical, chemical, biological, psychological, sociological, and epistemological parameters must all be in proper adjustment. There must be habitability, and life, and intelligence, and culture, and technology, and a science coupled to technology, and an appropriate subject-matter orientation of this intellectual product, etc. A great many turnings must go aright enroute to science of a quality comparable to ours. Each step along the way is one of finite (and often smallish) probability. To reach the final destination, all these probabilities must be multiplied together, yielding a quantity that might be very small indeed. Even if there were only twelve turning points along this developmental route, each involving a chance of successful eventuation that is, on average, no worse than a one in a hundred, the chance of an overall success would be immensely small, corresponding to an aggregate success probability of merely 10^{-24}.

It is tempting to say, "The universe is a big place; surely we must expect that what happens in one locality will be repeated someplace else," but this overlooks the issue of probability. Admittedly, cosmic locales are very numerous, but probabilities can get to be very small: No matter how massive N, there is that diminutive $1/N$ that can countervail against it.

The workings of evolution—be it of life or intelligence or culture or technology or science—are always the product of a great number of individually unlikely events. Things can eventuate very differently at many junctures. The unfolding of developments involves putting to nature a series of questions whose successive resolution produces a process reminiscent of the game "Twenty Questions," sweeping over a possibility spectrum of awesomely large proportions. The result eventually reached lies along a route that traces our one particular contingent path within a space of alternatives that provides for an ever-divergent fanning out of alternatives as each step opens up yet further possibilities. Too, the evolutionary process is a very iffy proposition—a complex labyrinth where a great many twists and turns in the road must be taken aright for matters to end up as they do.

Of course, it all looks easy with the wisdom of hindsight. If things had not turned out appropriately at every stage, we would not be here to tell the tale. The many contingencies on the long route of cosmic, galactic, solar-systemic, biochemical, biological, social, cultural, and cognitive evolution have all turned out aright—the innumerable obstacles have all been surmounted. In retrospect it all looks easy and inevitable. The innumerable possibilities of variation along the way are easily kept out of sight and out of mind. The wisdom of hindsight makes it all look very easy. It is so easy, so tempting to say that a planet on which there is life will of course evolve a species with the technical capacity for interstellar communication. (Cf. Cameron in Cameron, 1973, p. 312, who fixes this conditional probability as 1.) It is tempting, but it is also nonsense.

The ancient Greek atomists' theory of possibility affords an interesting object-lesson in this connection. Adopting a Euclideanly infinitistic view of space, the atomist taught that every (suitably general) possibility is realized in fact someplace or other. Confronting the question, "Why do dogs not have horns: just why is the theoretical possibility that dogs be horned not actually realized?" the atomists replied that it indeed is realized, but just elsewhere—*in another region of space.* Somewhere within infinite space there is another world just like ours in every respect save one, that its dogs have horns, for the circumstance that dogs lack horns is simply a parochial idiosyncrasy of the particular local world in which we interlocutors happen to find ourselves. Reality accommodates all possibilities of worlds alternative to this through spatial distribution: as the atomists saw it, *all* alternative possibilities are in fact actualized in the various subworlds embraced within one spatially infinite superworld.

This theory of virtually open-ended possibilities was shut off by the closed cosmos of the Aristotelian world picture, which dominated European cosmological thought for almost two millennia. The breakup of the Aristotelian model in the Renaissance and its replacement by the Newtonian model is one of the great turning points of the intellectual tradition of the West—elegantly portrayed in Alexandre Koyré's book of the splendid title *From the Closed World to the Infinite Universe* (New York: Knopf, 1957). Strangely enough, the refinitization of the universe effected by Einstein's general relativity produced scarcely a ripple in philosophical or theoretical circles, despite the immense stir caused by other aspects of the Einstein revolution. (Einsteinian space-time is, after all, even more radically finitistic than the Aristotelian world picture, which left open at any rate the prospect of an infinite future with respect to time.)

To be sure, it might well seem that the finitude in question is not terribly significant because the distances and times involved in modern cosmology are so enormous, but this view is rather naive. The difference between the finite and the infinite is as big as differences can get to be. It represents a difference that is—in this present context—of the most far-reaching significance. This means that we have no alternative to supposing that a highly improbable set of eventuations is not going to be realized in very many places and that something sufficiently improbable may well not be realized at all. The decisive *philosophical* importance of the Einsteinean finitization of space-time is that it means that an eventuation that is sufficiently improbable may well not be realized at all. A finite universe must make up its mind about its contents in a far more radical sense than an infinite one, and this is particularly manifest in the context of low-probability possibilities. In a finite world—unlike an infinite one—we cannot avoid supposing that a prospect that is sufficiently unlikely is simply not going to be realized at all, that in piling improbability on improbability we eventually outrun the reach of the actual. It is, accordingly, quite conceivable that our science represents a solution of the problem of cognitive accommodation that is terrestrially locale specific.

Here lies a deep question. Is the mission of intelligence uniform or diversified? Two fundamentally opposed philosophical positions are possible with respect to cognitive evolution in its cosmic perspective. The one is a uniformitarian *monism* that sees the universal mission of intelligence in terms of a certain shared destination, a common cosmic "position of reason as such." The other is a particularistic *pluralism* that allows each solar civilization to forge its own characteristic cognitive destiny and sees the mission of intelligence as such in terms of spanning a wide spectrum of alternatives and realizing a vastly diversified variety of possibilities, with each thought form realizing its own peculiar destiny in separation from all the rest. The conflict between these doctrines must in the final analysis be settled not by armchair speculation for general principles, but by rational triangulation from the empirical data. This said, it must be observed that the whole tendency of these present deliberations is toward the pluralistic side.

In many minds there is, no doubt, a certain charm to the idea of companionship. It would be comforting to think that however estranged we are in other ways, those alien minds and ourselves share *science* at any rate—that we are fellow travelers on a common journey of inquiry. Mythology and scientific speculation alike manifest our yearning for companionship and contact. (Pascal was not the only one frightened by the eternal silence of infinite spaces.) It would be pleasant to think ourselves not only colleagues but junior collaborators whom other, wiser minds might be able to help along the way. Even as many in sixteenth century Europe looked to those strange pure men of the Indies (East or West) who might serve as moral exemplars for sinful European man, so are we tempted to look to alien inquirers who surpass us in scientific wisdom and might assist us in overcoming our cognitive deficiencies. The idea is appealing, but it is also, alas, very unrealistic.

In the late 1600s Christiaan Huygens wrote:

> For 'tis a very ridiculous opinion that the common people have got among them, that it is impossible a rational Soul should dwell in any other shape than ours . . . This can proceed from nothing but the Weakness, Ignorance, and Prejudice of Men, as well as the humane Figure being the handsomest and most excellent of all others, when indeed it's nothing but a being accustomed to that figure that makes me think so, and a conceit . . . that no shape or color can be so good as our own. (Huygens, 1698, pp. 76-77.)

What is here said about people's tendency to emplace all rational minds into a physical structure akin to their own familiar one is paralleled by a tendency to emplace all rational knowledge into a cognitive structure akin to their own familiar one.

With respect to biological evolution, it seems perfectly sensible to reason as follows:

> What can we say about the forms of life evolving on these other worlds? . . . [I]t is clear that subsequent evolution by natural selection would lead to an immense variety of organisms; compared to them, all organisms on Earth, from molds to men, are very close relations. (Shklovskii and Sagan, 1966, p. 350).

It is plausible that much the same situation should obtain with respect to cognitive evolution: that the sciences produced by different civilizations here on earth—the ancient Chinese, Indians, and Greeks for example—should exhibit immensely greater points of similarity than obtains between our present-day science and anything devised by astronomically remote civilizations. The idea of a comparison in terms of "advance" or "backwardness" would simply be inapplicable.

This chapter's line of reflection conveys two principal lessons. The first is that the prospect that some astronomically remote civilization is scientifically more advanced than ourselves—that somebody else is doing our sort of science *better* than we ourselves—requires in the first instance that they be doing our sort of science at all. This deeply anthropomorphic supposition is extremely unlikely. A second main lesson follows from the consideration that natural science *as we know it* is to all visible intents and purposes a characteristically human enterprise—a circumstance that endows science with an inexorably economic dimension. This means that the sorts of results of scientific inquiry that we are able to achieve will hinge crucially on *the way* in which we deploy resources in cultivating our scientific work as well as on *the extent* to which we do so.[1]

6. SELECT BIBLIOGRAPHY

NOTE: This listing is confined to those relevant materials that I have found particularly interesting or useful. It does not aspire to comprehensiveness. A much fuller bibliography is given in MacGowan and Ordway, 1966.

Allen, Thomas Barton. *The Quest: A Report on Extraterrestrial Life*. Philadelphia: Chilton Books, 1965. (An imaginative survey of the issues.)

Anderson, Paul. *Is There Life on Other Worlds?* New York and London: Collier-MacMillan, 1963. (Chapter 8, "On the Nature and Origin of Science," affords many perceptive observations.)

Ball, John A. "The Zoo Hypothesis," *Icarus* 19 (1973): 347-349. (Aliens are absent because the Intergalactic Council has designated Earth a nature reserve.)

_____. "Extraterrestrial Intelligence: Where Is Everybody?" *American Scientist* 68 (1980): 556-663.

Beck, Lewis White. "Extraterrestrial Intelligent Life," *Proceedings and Addresses of the American Philosophical Association* 45 (1971-72): 5-21. (A thoughtful and very learned discussion.)

Berrill, N. J. *Worlds without End*. London: MacMillan, 1964. (A popular treatment.)

Bracewell, Ronald N. *The Galactic Club: Intelligent Life in Outer Space*. San Francisco: W. H. Freeman, 1975. (A lively and enthusiastic survey of the issues.)

Breuer, Reinhard. *Contact with the Stars*. Trans. C. Payne-Gasposchkin and M. Lowery. New York: W. H. Freeman, 1982. (Maintains that we are the only technologically developed civilization in the galaxy.)

Cameron, A. G. W., ed. *Interstellar Communication: A Collection of Reprints and Original Contributions.* New York and Amsterdam: W. A. Benjamin, 1963. (A now somewhat dated but still useful collection.)

Dick, Steven J. *Plurality of Worlds: The Origins of the Extraterrestrial Life Debate from Democritus to Kant.* Cambridge: Cambridge University Press, 1982. (A lively and informative survey of the historical background.)

Dole, Stephen H. *Habitable Planets for Man.* New York: Blaisdell, 1964; 2nd. ed., New York: American Elsevier, 1970. (A painstaking and sophisticated discussion.) A more popular version is S. H. Dole and Isaac Asimov, *Planets for Man.* New York: Random House, 1964.

Drake, Frank D. *Intelligent Life in Space.* New York and London: Macmillan, 1962. (A clearly written, popular account.)

Ehrensvaerd, Goesta. *Man on Another World.* Chicago and London: University of Chicago Press, 1965. (See especially chapter 10 on "Advanced Consciousness.")

Firsoff, M. V. A. *Life beyond the Earth: A Study in Exobiology.* New York: Basic Books, 1963. (A detailed study of the biochemical possibilities for extraterrestrial life.)

Gabbay, Allen. "Les Principes foundamenteaux de la connaissance: Le Modele des intelligences extraterrêtres." *Science, Histoire, Épistémologie*: Actes du Premier Colloque Européen d'Histoire et Philosophie des Sciences. Paris: J. Vrin, 1981, 33-59. (A stimulating philosophical discussion.)

Hart, M. H. "An Explanation for the Absence of Extraterrestrials on Earth," *Quarterly Journal of the Royal Astronomical Society* 16 (1975): 128-135. (A perceptive survey of this question.)

Herrmann, Joachim. *Leben auf anderen Sternen.* Guetersloh: Bertelsmann Verlag, 1963. (A thoughtful and comprehensive survey with special focus on the astronomical issues.)

von Hoerner, Sebastian. "Astronomical Aspects of Interstellar Communication," *Astronautica Acta* 18 (1973): 421-429. (A useful overview of key issues.)

Hoyle, Fred. *Of Men and Galaxies*. Seattle: University of Washington Press, 1966. (Speculations by one of the leading astrophysicists of the day.)

Huang, Su-Shu. "Life outside the Solar System," *Scientific American* 202, 4 (April 1960): 55-63. (A useful discussion of some of the astrophysical issues.)

Huygens, Christiaan. *Cosmotheoros: The Celestial Worlds Discovered—New Conjectures Concerning the Planetary Worlds, Their Inhabitants and Productions*. London, 1698; reprinted London: F. Cass & Co., 1968. (A classic from another age.) Cf. Dick, 1982.

Jeans, Sir James. "Is There Life in Other Worlds?" A 1941 Royal Institution lecture reprinted in H. Shapley et al., eds., *Reading in the Physical Sciences*. New York: Apple-Century-Crofts, 1948, 112-117. (A stimulating analysis.)

Kaplan, S. A., ed. *Extraterrestrial Civilization: Problems of Interstellar Communication*. Jerusalem: Israel Program for Scientific Translations, 1971. (A collection of Russian scientific papers that present interesting theoretical work.)

Lem, S. *Summa Technologiae*. Krakow: Wyd. Lt., 1964. (To judge from the ample account given in Kaplan, 1971, this book contains an extremely perceptive treatment of theoretical issues regarding extraterrestrial civilizations. I have not, however, been able to consult the book itself.)

MacGowan, Roger A., and Ordway, Frederick I., III. *Intelligence in the Universe*. Englewood Cliffs, N.J.: Prentice Hall, 1966. (A careful and informative survey of a wide range of relevant issues.)

_____. "On the Possibilities of the Existence of Extraterrestrial Intelligence." In F. I. Ordway, ed., *Advances in Space Science and Technology*. New York and London: Academic Press, 1962. 4:39-111.

Nozick, Robert. "R.S.V.P.—A Story," *Commentary* 53 (1972): 66-68. (Perhaps letting aliens know about us is just too dangerous.)

Pucetti, Roland. *Persons: A Study of Possible Moral Agents in the Universe.* New York: Herder and Herder, 1969. (A stimulating philosophical treatment.) But see the sharply critical review by Ernan McMullin in *Icarus* 14 (1971): 291-294.

Rood, Robert T., and Trefil, James S. *Are We Alone: The Possibility of Extraterrestrial Civilization.* New York: Scribners, 1981. (An interesting discussion of the key issues.)

Sagan, Carl. *The Cosmic Connection.* New York: Doubleday, 1973. (A well-written, popularly oriented account.)

_____. *Cosmos.* New York: Random House, 1980. (A modern classic.)

Shapley, Harlow. *Of Stars and Men.* Boston: Beacon Press, 1958. (See especially the chapter entitled "An Inquiry Concerning Other Worlds.")

Shklovskii, I. S., and Sagan, Carl. *Intelligent Life in the Universe.* San Francisco, London, Amsterdam: Holden-Day, 1966. (A well-informed and provocative survey of the issues.)

Simpson, George Gaylord. "The Nonprevalence of Humanoids," *Science* 143 (1964): 769-775, chapter 13 of *This View of Life: The World of an Evolutionist.* New York: Harcourt Brace, 1964. (An insightful account of the contingencies of evolutionary development by a master of the subject.)

Sullivan, Walter. *We Are Not Alone.* New York: McGraw Hill, 1964, rev. ed. 1965. (A very well-written survey of the historical background and of the scientific issues.)

NOTES

1. This chapter draws upon the author's essay "Extraterrestrial Science," *Philosophia Naturalis*, vol. 21 (1984), pp. 400-424.

Chapter Nine

PROBLEMS OF
SCIENTIFIC REALISM

SYNOPSIS

(1) Scientific realism is the doctrine that science describes the real world. This is certainly not the case with our science as it stands: we are clearly not entitled to claim that science as we have it is either definitively correct or fully complete, free of all errors of commission or omission. (2) Nor can we plausibly make this sort of claim on behalf of future science. The nature of natural science as a human productive enterprise means that the project at issue is inherently incompletable and that the disabilities of present science cannot be eliminated altogether at any particular point in the future. (3) The most that can be claimed in the direction of scientific realism is that it is ideal science—and not science as we have it—that correctly describes the real world. The tentative nature of science as we have it, now or ever, means that the only tenable form of scientific realism is its ideal-science version.

1. SCIENTIFIC REALISM

Scientific realism is the doctrine that *science describes the real world*—that the world actually is as science takes it to be and that its furnishings are as science envisions them to be.[1] Accordingly, scientific realism maintains that such theoretical entities as the quarks and electrons of contemporary science are perfectly real components of nature's real world, every bit as real as acorns and grains of sand. The latter we observe with the naked eye, the former we detect by complex theoretical triangulation, but a scientific realism of theoretical entities maintains that this difference is incidental. In principle, these unobservable entities exist in just the way in which the scientific theories that project them maintain. On such a realistic construction of scientific theorizing, the declarations of science are factually true generalizations about the actual behavior of real physical objects existing in nature.

But is this a tenable position? Clearly, it has difficulties, for the theoretical entities envisioned by current science will exist as current science envisions them only insofar as current science is in fact correct—only if it manages to get things just right. The characteristic genius of scientific realism is inherent in its equating of the theory creatures envisioned in current natural science with the domain of what actually exists. This equation would work only if our science, as it stands here and now, has actually got it right. This is something we are certainly not to claim, for we know full well that science constantly changes its mind, not just with regard to incidentals but even on very fundamental issues.

The history of science is the story of the replacement of one defective theory by another. We realize full well that the scientists of the year 3000 will think of our science as correct no more than we so think of the science of 100 years ago. So how can one plausibly maintain a scientific realism geared to the idea that science correctly describes reality? All too clearly there is insufficient warrant for and little plausibility to the claim that the world is as our present-day science claims it to be—that *our* science is *correct* science and offers the definitive last word on the issues regarding its creatures of theory. We can learn by empirical inquiry about empirical inquiry itself, and one of the key things to be learned is that at no actual stage does natural science yield a firm, final, unchanging result.

The ultimate untenability of our scientific theories is in fact one of the very few points of consensus of modern philosophy. When Karl Popper writes, "From a rational point of view, we should not 'rely' on any (scientific) theory, for no theory has been shown to be true, or can be shown to be true . . . "[2], he speaks for the entire tradition of modern science scholarship from Charles Sanders Peirce to Nancy Cartwright. We must unhesitatingly presume that, as we manage to push our inquiries through to deeper levels of understanding, we will get a very different view of the constituents of nature and their laws. Its changeability is a fact *about* science that is as inductively well established as any theory *of* science itself. Science is not a static system but a dynamic process. The constant progress of science is a matter of constant changes of mind.

Postulating the reality of science's commitments is viable only if done *provisionally*, in the spirit of doing the best we can manage at present, in the current state of the art. Experience teaches that our prized scientific knowledge is no more than our current best estimate of the matter. The step of reification is always to be taken provisionally, subject to a mental reservation of presumptive revisability.

We may *think* we are bound to have encompassed the truth when we have boxed the compass of alternatives:

—Cancer is caused by a virus.

—Cancer is caused by something other than a virus.

Surely, we tell ourselves, the truth must lie either on the one side or the other. The *disjunction* of these possibilities must evidently be true.

But is it really so? Observe that both theses alike adopt a common presupposition. Both envision a disease entity designated by the term *cancer*, but what if there is no such *thing* as "cancer" because there just is not a stable entity, but merely a diversified family of ailments united only by Wittgensteinian family resemblances? What if we come to believe that the very concept at issue is not objectively meaningful—if cancer as such simply is not there to have a cause at all. It is a presupposition of every factual statement that its concepts have a bearing on the real world, that they are indeed applicable to things, that nature actually exemplifies them, that they are "objectively meaningful." As "magnetic effluxes" and the "luminiferous aether" show, however, this presupposition can be totally mistaken in scientific contexts.

The current state of scientific knowledge is simply one state among others that share the same imperfect footing as regards ultimate correctness or definitive truth. The science of the day must be presumed inaccurate no matter what the calendar says. We unequivocally realize there is a strong prospect that we shall ultimately recognize many or most of our current scientific theories to be false and that what we proudly vaunt as scientific knowledge is a tissue of hypotheses —of tentatively adopted contentions, many or most of which we will ultimately come to regard needing serious revision or perhaps even abandonment.

We must therefore maintain a certain tentative and provisional stance toward our own scientific knowledge. We fully realize that what we *take* to be true or real here is not always true or real. It is just this consideration that constrains us to operate with the distinction between our putative reality and reality as such. We realize that what we think to be so—be it in science or in common life—frequently just is not so.

We learn by empirical inquiry about empirical inquiry, and one of the salient things we learn is that at no actual stage does science yield a final and unchanging result. All the experience we can muster indicates that there is no justification for regarding our present-day science as more than an inherently imperfect stage within an ongoing development. The landscape of science is ever changing.

This state of affairs blocks the option of a scientific realism of any straightforward sort. Not only are we not in a position to claim that our knowledge of reality is *complete* (that we have gotten at the *whole* truth of things), but we are not even in a position to claim that our knowledge of reality is *correct* (that we have gotten at the *real* truth of things). Such a position calls for the humbling view that just as we think our predecessors of a hundred years ago had a fundamentally inadequate grasp on the furniture of the world, so our successors of a hundred years hence will take the same view of our purported knowledge of things.

Thus, a clear distinction must be maintained between *our conception of* reality and reality *as it really is.* We realize that there is precious little justification for holding that present-day natural science describes reality and depicts the world as it really is. This constitutes a decisive impediment to any straightforward realism. It must inevitably constrain and condition our attitude toward the natural mechanisms envisioned in contemporary science. We certainly do not—or should not—want to reify (hypostatize) the theoretical entities of current science, to say flatly and without qualification that the contrivances of *our* present-day science correctly depicts the furniture of the real world. We do not—or at any rate, given the realities of the case, should not—want to adopt categorically the ontological implications of scientific theorizing in just exactly the state of the art configuration presently in hand. A realistic acknowledgment of scientific fallibilism precludes the claim that the furnishings of the real world are exactly as our science states them to be—that electrons actually are just what the latest *Handbook of Physics* claims them to be.

Committed to the unproblematic claim *that* reality exists, we are, nevertheless, equally committed to the supposition that its nature is, in various not unimportant ways, different from what we think it to be. We can make no assured claims for our present-day science in this matter of describing reality: the most we can do is to see it as affording our very best estimate of nature's descriptive constitution. We realize that science as it stands does not give us definitive knowledge. We know that we will eventually come to see with the wisdom of hindsight that each of the claims of current frontier science, taken literally in the fullness of current understandings and explanations, is, strictly speaking, false.[3] The realities of the situation force us to accept the presumptive falsity of the claims made at the scientific frontier of the present day.

Scientific progress is *not* of a character that encourages us to reify (hypostatize) the theory objects of science *as presently conceived*—regardless of the date the calendar may show. Once we have taken a realistic look at the history of science, it is scarcely a plausible proposition to maintain that *our* science, as it stands here and now, depicts reality actually and correctly— at best one can say that it affords an *estimate* of it that will doubtless stand in need of eventual revision. Its creatures of theory may in the final analysis not be real at all in the form in which the theory envisions them. This feature of science must crucially constrain our attitude toward its deliverances.

We have to come to terms with the realism-impending fact that our scientific knowledge of the world fails in crucial respects to give an accurate picture of it. Certainly, we subscribe for the most part to the working hypothesis that in the domain of factual inquiry *our* truth may be taken to be *the* truth. All the same, we realize that our science is not definitive, that reality is *not* actually as we currently picture it to be, that our truth is not the real truth, that we are probably quite wrong in supposing that the furnishings of our science actually exist exactly as it conceives them to be. No doubt, reality itself, whatever that may be, stands secure, but our empirical reality—reality as our science conceives it—wears the hallmark of a plausible fiction. Ultimately, when science is seen in a historical perspective as the ongoing process it is, it becomes clear that there is no adequate justification for thinking that natural science as we now have it manages to get it right about things of the right sort.

2. The Imperfections of Future Science

The crucial fact is that our material resources are limited and these limits inexorably circumscribe our cognitive access to the real world. There are interactions with nature of so massive a scale (as measured in such parameters as energy, pressure, temperature, particle velocities, etc.) that their realization would require the deployment of resources so great that we can never achieve them. If, however, there are interactions to which we have no access, then there are (presumably) phenomena that we cannot discern. It would be unreasonable to expect nature to confine the distribution of phenomena of potential cognitive significance to those ranges that lie within the horizons of our present vision.

Ultimately, there will then always be interactions with nature on a scale whose realization requires the deployment of greater resources than we have heretofore expended. Humanity's material resources are limited, and these limits inexorably circumscribe our experiential access to the real world. When there are interactions to which we have no access, we must presume phenomena that we cannot discern. It would be very odd indeed if nature were to confine the distribution of scientifically significant phenomena to those ranges that happen to lie conveniently within our reach—a condition counterindicated by the whole course of our prior experience.

Given that economic resource limitations will, on this basis, ultimately limit our access to nature's phenomena, we must come to terms with the fact that we cannot realistically expect that our science will ever, at any given stage of its development, be in a position to provide more than a partial and incomplete account of nature. It will never attain the stabilization of a final completion, for the achievement of cognitive control over nature requires not only intellectual instrumentalities (concepts, ideas, theories, knowledge) but also, and no less importantly, the deployment of physical ones (technology and power). Since the physical resources at our disposal are restricted and finite, it follows that our capacity to effect control is bound to remain imperfect and incomplete, with much in the realm of the doable always remaining undone.

Thus, while we can confidently anticipate that our scientific understanding will see ongoing improvement, we cannot expect it ever to attain perfection. There is no reason to think that we ever will, or indeed can, reach the end of the line. Every successive level of technical capability has its inherent limits, whose overcoming opens up yet another more sophisticated level of the technological state of the art whose ultimate fate is in its turn of the same nature. The accessible pressures and temperatures can in theory always be increased, the low-temperature experiments brought closer to absolute zero, the particles accelerated closer to the speed of light, and so on. Experience teaches, too, that any such enhancement of practical mastery carries along new phenomena, engendering an enhanced capability to test yet further hypotheses and discriminate between alternative theories conducive to deepening our knowledge of nature. There is always more to be done.

The material resources available for scientific work are inevitably limited, however, and these limits inexorably circumscribe our cognitive access to the real world. Advancing natural science in any given direction requires deploying some appropriate mode of data technology. If and when a point is reached where technology improvement becomes unaffordable—or impractical for any other reason—the relevant branch of science is condemned to stagnation. The fact of the matter is, however, that there will always be interactions with nature of such a scope that their realization would require the concurrent deployment of resources so vast that their realization in unachievable for us. When there are interactions to which we have no access, then there are always phenomena that we cannot discern. (It would be very unreasonable—and entirely unwarranted—to expect nature to confine the distribution of cognitively significant phenomena to those ranges that lie within our reach at any given time, and in fact all the evidential indications afforded by history point the other way.)

In scientific contexts, limited physical control means limited cognitive control. The layering of our technological capacity at successive state-of-the-art stages reflects a *sequencing* in the sorts of discoveries that are available. One cannot do subatomic physics until atomic physics has been developed; one cannot progress in immunology until bacteriology has been developed. One must round the earlier turns in the road of discovery before the later can be reached. This means that some fundamental aspects of nature are hidden from view in any and every particular state of the scientific art.

We must not delude ourselves into thinking that physical control does not matter for understanding that *praxis* is irrelevant to *theoria*. Natural science is *empirical* science. Theory is intimately bound up with experimentation and observation. One of the clearest lessons of the history of science is that as we acquire more powerful means for data acquisition and processing, and thus as our information base changes, so the character of our theories, and with it our view of the world, our "picture of nature," also changes. The existence of a potentially unending sequence of levels of *technological* sophistication entails an unending sequence of levels of *theoretical* sophistication, with a very different story, a different picture of nature, emerging at every level. But it gets increasingly expensive—and ultimately, in a world of limited resources, simply too expensive—to advance in the frontiers of theorizing. We must, accordingly, come to terms with the fact that we cannot realistically expect that our science will ever—at *any* given stage of its actual development—be in a position to afford us more than a very partial and incomplete physical control over nature.

Once we distance ourselves from the cognitive commitments of our science by recognizing that they can and frequently do go awry, we must also acknowledge that "our scientific picture" of reality is not fully accurate, recognizing that we lack the warrant for claiming that reality actually is as it is purported to be by the science of the day.   In the context of our cognitive endeavors, "man proposes and nature disposes" and it does so in both senses of the term: it disposes *over* our current scientific view of reality and it will doubtless eventually dispose *of* it as well.   Given this circumstance, we have little alternative but to presume reality to have a character regarding which we are only imperfectly informed by natural science.   This holds as much for the science of the future as it does for the science of the past.

Limitations of physical capacity and capability inexorably produce *cognitive* limitations for empirical science.   Where there are inaccessible phenomena, there must be cognitive inadequacy as well.   To this extent, at any rate, traditional empiricists were surely right: sheer intellect alone cannot compensate for the lack of data.   The existence of unobserved phenomena means that our theoretical systematizations may well be (and presumably are) incomplete.   Insofar as certain phenomena are not just undetected but in the very nature of the case inaccessible (even if only for the merely economic reasons suggested above), our theoretical knowledge of nature must be presumed imperfect.   The nature of natural science as a human productive enterprise thus means that the project at issue is inherently incompletable and that the disabilities of present science cannot be eliminated altogether at any particular point in the future.   In science, as elsewhere, perfection is something we cannot afford— now or ever.

3. THE ASPECT OF IDEALIZATION

Fundamental features inherent in the very structure of human inquiry into the world's ways conspire to ensure the incompleteness and imperfection of the knowledge we can attain in this sphere. The practical and economic realities of science's development lead back to the thesis of the great idealist philosophers (Spinoza, Hegel, Bradley, Royce) that human knowledge inevitably falls short of perfected science (the Idea, the Absolute) and must be presumed deficient both in its completeness and its correctness.

Recognizing the facts of life in empirical inquiry, we have neither the inclination nor the justification to claim that the world is as our *present* science describes it to be. Nor, as we have seen, does it make sense to identify the real truth with the truth as science in the limit will eventually see it to be. The best that can be done in this direction is to say that the world exists as *ideal* or *perfected* science describes it to be. The real, which is to say final and definitive, truth about nature at the level of scientific generality and precision is something we certainly cannot assume *our* science to capture. We have no choice but to take the stance that it is not something we actually have, but something that—so we must suppose—is attained only in the ideal or perfected state of things. We thus arrive at the equation:

the real truth = the truth as ideal (perfected) science purports it to be.

We certainly cannot identify our achieved putative scientific truth with the real truth of the matter. No route save idealization is able to effect a sure and general connection between belief and the real truth. Only ideal or perfected science accurately and correctly depicts reality, and not science as we actually have it here and now. From the standpoint of epistemic status, truth is clearly an idealization—not what we *do* (or even *will*) *have*, but what we *could have if* all the returns were in. It is thus in order to take a closer look at this matter of cognitive idealization.

Success in providing a definitive truth about nature's ways is doubtless a matter of intent rather than one of accomplishment. Correctness in the characterization of nature is achieved not by *our* science, but only by *perfected* or *ideal* science—only by that (ineradicably hypothetical) state of science in which the cognitive goals of the scientific enterprise are fully and reliably realized because the limitation of resources—material or intellectual—stand in the way. We are constrained to acknowledge that it is not *present* science, not even *future* or *ultimate* science, but only *ideal* science that correctly describes reality—an ideal science that we shall never in fact attain, since it exists only in utopia and not in this mundane dispensation. Scientific realism must thus come to terms with the realization that reality is depicted by *ideal* (or perfected or completed) science, and not by the real science of the day, which, after all, is the only one we have actually got—now or ever.

Of course, the concept of science perfected—of an ideal and completed science that captures the real truth of things and satisfies all of our cognitive ideals (definitiveness, completeness, unity, consistency, etc.) —is at best a useful fiction, a creature of the imagination and not the secured product of inquiring reason. This ideal science that attains definitive finality is, as the very name suggests, something of a pie in the sky. It represents an idealization and not a matter of the practical politics of the epistemic domain, but it affords the *focus imaginarius* whose pursuit canalizes and structures our inquiry. It represents the ultimate goal of inquiry—the destination of a journey in which we are *still* and indeed are *ever* engaged.

The conception of definitive capital-T Truth in matters of natural science thus serves a negative and fundamentally regulative role to mark the fact that the place we have attained falls short of our capacity actually to realize our cognitive aspirations. It marks a fundamental contrast that *regulates* how we do and must view our claims to have gotten at the truth of things. It plays a rule somewhat reminiscent of the functionary who reminded the Roman emperor of his mortality in reminding us that our pretensions to truth are always vulnerable. Contemplation of this ideal enables us to maintain the ever-renewed recognition of the essential ambiguity of the human condition as suspended between the reality of imperfect achievement and the ideal of an unattainable perfection. In abandoning this conception—in rejecting the idea of an ideal science that alone can properly be claimed to afford a grasp of reality—we would abandon an idea that crucially regulates our view as to the nature and status of the knowledge we lay claim to. We would then no longer be constrained to characterize our putative truth as *merely* ostensible and purported, and then, if our truth did not exhibit any blatant *inherent* imperfections, we would be tempted to view it as real, authentic, and final in a manner that as we at bottom realize it does not deserve. We must presume that science cannot attain an omega-condition of final perfection. The prospect of fundamental changes lying just around the corner can never be eliminated finally and decisively. However confidently science may affirm its conclusions, its declarations are effectively provisional and tentative, subject to revision and even to outright abandonment and replacement.

Now, ideal science is not something we have got in hand here and now, nor is it something toward which we are moving along the asymptotic and approximative lines envisioned by Charles Sanders Peirce,[4] who identified it with an ultimate condition of science that is "fated" to emerge in the eventual course of history, for there is, of course, no guarantee of this whatsoever. Perfected science is not "what will emerge when" but "what would emerge if"—where a lot of (realistically unachievable) conditions must be supplied. As far as the actual course of history goes, we must recognize that even if it made sense to contemplate the Peircean idea of an eventual completion of science, there would be no guarantee that this completed science (given it existed!) would satisfy the definitive requirements of *perfected science*. Peircean convergentism is geared to the supposition that ultimate science—the science of the very distant future—will somehow prove to be an ideal or perfected science freed from the sorts of imperfections that afflict its predecessors. The potential gap that arises here can be closed, however, only by metaphysical assumptions of a most problematic sort.

We recognize full well that our scientific inquiries about how things work in the world do not provide us with the real (definitive) truth, but rather with *the best estimate* of the truth we can achieve in the circumstances to hand. Our science is constituted of *putative* knowledge that does no more than envision the truth as best we can discern it with the limited means at our disposal. Only at the idealized level of perfected science could we count on securing the real truth about the world that "corresponds to reality," as the traditional phrase has it.

Existing science does not and presumably never will embody to perfection cognitive ideals of definitiveness, completeness, unity, consistency, etc. These factors represent an aspiration rather than a coming reality: a *telos* or direction rather than a realizable condition of things. Accordingly, there is no warrant for identifying *ideal* or perfected science with *ultimate* science. Perfected science is not something that exists here and now, nor is it something that lies ahead at some eventual offering in the remote future. It is not a real thing to be met with in this world. It is an idealization that exists "outside time"—i.e., cannot attain actual existence at all. It lies outside history as a useful contrast case that cannot be numbered among the achieved realities of this imperfect world.

The concept of science perfected—of an ideal and completed science that captures the real truth of things and satisfies all of our cognitive ideals (definitiveness, completeness, unity, consistency, etc.)—is at best a useful fiction, a creature of the fictive imagination and not the secured product of inquiring reason. This "ideal science" is, as the very name suggests, an idealization. We can only do the best we can in the cognitive state of the art to *estimate* "the correct" answer to our scientific questions, which must suffice us because it is *all* that we can do.

Scientific inquiry is truth-estimation, and here, as elsewhere, the gap between the real and the ideal must be acknowledged. What inquiry provides is "our purported truth" as contradistinguishable from "the real truth itself." The idea of "the definitive truth" functions as a regulative conception for us. It characterizes what we ideally aim at rather than describing its achievements.[5]

We recognize, or at any rate have little alternative but to suppose, that reality exists, accepting that there is such a thing as "the real truth" about the mind-independently real things of this world, but we are not in a position to state any final and definitive claims as to just exactly what it is like. Here we are confined to the level of plausible conjecture—of estimation.

One recent commentator maintains the view that science's aim regarding true theories "leads to the view that science represents a utopian, and therefore irrational activity whose *telos* is, to the best of our knowledge, forever beyond our grasp."[6] This position is profoundly wrong. It fails to deal appropriately with the standard gap between aspiration and attainment. In the practical sphere—in our craftsmanship, for example, or our health care—we may *strive* for perfection, but cannot ever claim to have *attained* it. The situation of inquiry is exactly parallel with what we encounter in other domains—ethics specifically included. The value of a goal, even of one that is not realizable, lies not in the benefits of its attainment (obviously and *ex hypothesi*!), but in the benefits that accrue from its pursuit. Even an unattainable goal can be perfectly valid and entirely rational if the indirect benefits of its adoption and pursuit are sufficient—if in striving after it we realize relevant advantages to a substantial degree. Such "unrealistic" objectives can in fact be enormously productive.

There is only one world in existence: the real world as it actually is. We will not be able to say just what it is really like until the day when natural science has been completed and perfected, which is to say *never*. The practical and economic realities of the situation mean that we must pursue the cognitive enterprise amid the harsh realities and complexities of an imperfect world. This in turn means that what we achieve in scientific inquiry is not the definitive truth as such, but only our best estimate of it. In forming a just appreciation of our scientific claims, we must come to terms once again with the irremovable gap between the real and the ideal.

The thesis that science truly describes the real world must be looked upon as a matter of intent rather than as an accomplished fact, of aspiration rather than achievement, of the ideal rather than the real state of things. Scientific realism is therefore tenable only when it is the *ideal* state of science that is at issue—a science unlimited by resource scarcity of any sort. This ideal-state realism, however, while no doubt correct, avails us less than we would like—we who occupy the sub-optimally real rather than the perfected ideal order of things. Science as we actually have it—now or ever—is and can be no more than the best estimate of the real condition of things that we can currently make with the nature-interactive resources at our disposal. For this estimate we will never be able to claim either comprehensive completeness or definitive correctness.[7]

NOTES

1. Ontological realism contrasts with ontological *idealism*; scientific realism contrasts with scientific *instrumentalism*: the doctrine that science in no way describes reality, but merely affords a useful organon of prediction and control.

2. Karl R. Popper, *Objective Knowledge* (Oxford: Clarendon Press, 1972), p. 9.

3. This, in effect, is the salient insight of twentieth century philosophy of science from C. S. Peirce through K. R. Popper's *Logik der Forschung* (Tübingen: Mohr, 1934) to Nancy Cartwright's *How the Laws of Physics Lie* (New York: Oxford University Press, 1984).

4. *Cf.* the author's *Peirce's Philosophy of Science* (Notre Dame and London: University of Notre Dame Press, 1978).

5. This is not, of course, any reason to abandon the link to truth at the purposive level of the aims, goals, and aspirations of science. The pursuit of scientific truth, or, for that matter, any other ideal in life, is not vitiated by the consideration that its full realization is not achievable.

6. Larry Laudan, "The Philosophy of Progress," mimeographed preprint (Pittsburgh, 1979), p. 4. *Cf. idem, Progress and its Problems* (Berkeley and Los Angeles: University of California Press, 1977).

7. Some of the themes of this chapter are also touched upon in Chapters 5 and 7 of the author's *Empirical Inquiry* (Totowa, N.J.: Rowman & Littlefield, 1982).

CONCLUSION

In concluding, it is fitting to pass in review the main contentions of the preceding deliberations. They are as follows:

- Natural science is not a substantial thing of some sort, a "body of knowledge"; it is a living, changing, developing process, correlative with the activity of inquiring that represents an ongoing reciprocal interaction between ourselves and nature.

- There are no inherent limits to this process of cognitive development. The idea of bringing the process of scientific inquiry to a conclusion is altogether impracticable. There is no good reason to think that it will ever terminate in an inevitable end of some sort.

- Nor can we justify the idea that the process of scientific inquiry yields a product that converges asymptotically toward some fixed and stable condition of things.

- Scientific inquiry as a human activity does, however, have unavoidable economic involvements, as any activity must. It is, in fact, deeply—and inseparably—interconnected with the process of technological development.

- The economic aspect of science means that, at any particular stage in the state of the art of data acquisition and exploration, there is only so much science that we can afford.

- For a whole group of interrelated reasons, theoretical and practical alike, we can never advance the project of scientific inquiry as far as we would like and as far as is theoretically possible. Our science will always manifest errors of omission and errors of commission. The idea of a perfected science is pie in the sky.

- Science as we humans can and do actually develop it is bound to reflect the way in which we humans are inserted through evolution within our actual and contingent environment. Science as we know it is a characteristically human product whose nature depends on the particular way in which we invest resources in it.

- This means that the philosophical doctrine of scientific realism, which maintains that reality actually and definitively is as science teaches it to be, is unavoidably involved in a process of idealization. It simply cannot be implemented with respect to real (as opposed to ideal) science.

The upshot of these deliberations is that there is nothing definitive and absolute about natural science as we do or can have it. Science is a human artifact—the product of our cognitive endeavor in relation to man-nature indications. And, as with any other human artifice the materials made available to us by nature make for limits and limitations, providing restrictions to what we can do. There is an ineliminable economic dimension to natural science because the increasingly sophisticated manipulation of physical materials is an indispensable requisite. Scientific information may be priceless in value, but its obtaining always and inevitably involves a genuine and substantial cost. The crucial fact is that here as elsewhere one must pay for quality—and that improvements in product are available to us only at an ever increasing price. In the case of science this price will ultimately rise to a level that we shall find effectively impossible to pay.

One hears it argued that science is incomplete because it does not achieve our normative—and in particular our moral and aesthetic—goals and aspirations. In this regard the present discussion takes a very different line in arguing that it does not ever fully satisfy our *cognitive* aspirations. It represents no more—but also no less—than the imperfect effort by imperfect humans to do the best they can in this direction.

All the same, an important final point deserves stress. Even if it were certain that mankind had pushed its effort at scientific inquiry deep into the region of diminishing returns, one should not suppose that this had gone too far. The evaluation of the *costs* of scientific knowledge in terms of material and intellectual resources must be offset by recognition of the *benefits* of scientific work, and these benefits should not be construed in the narrowly utilitarian, applications-oriented sense. It is important to recognize that not just *material* but *intellectual* benefits are involved, for while scientific progress has indeed produced an immense benefit in terms of physical well-being, there remains the no-less-crucial fact that it represents one of the great creative challenges of the human spirit.

Our intellectual struggle with nature deserves to be ranked as a key element of what is truly *noble* in human life, together with our social efforts at forging a satisfying life environment and our moral strivings to transcend the limitation of our animal heritage. A society that spends many billions of dollars on a varied cornucopia of deleterious trivia, to say nothing of untold billions on military outlays and mind-numbing diversions, assumes an uncomfortable moral posture in deciding that science—even big and expensive science—is a game that's not worth the candle.[1] The scientifically and technologically most advanced countries nowadays spend some 2-3 percent of their GNP on research and development and some 3 percent of that on basic science.[2] This allocation of roughly onetenth of one percent of GNP to pure science is certainly not exorbitant —perhaps not even seemly, considering the size of our material and intellectual stake in the enterprise.[3]

NOTES

1. The moral aspect aside, it is probably not even a matter of prudentially enlightened self-interest. Compare the cogent discussion in Stephen Toulmin, "Is There a Limit to Scientific Growth?" *Scientific Journal*, vol. 2 (1966), pp. 80-85.
2. See Keith Norris and John Vaizey, *The Economics of Research and Technology* (London: Allen & Unwin, 1973), p. 56. The situation has deteriorated over the quarter-century since this study was made.
3. The themes at issue in these closing remarks are further explored in the author's *Scientific Progress* (Oxford: Basil Blackwell, 1978).

NAME INDEX

SUBJECT INDEX

About the Author

Nicholas Rescher was born in Hagen, Germany, in 1928 and came to the United States at the age of ten. He attended Queens College in New York City and Princeton University, where he earned his Ph.D. at age twenty-two. Since 1961 he has been University Professor of Philosophy at the University of Pittsburgh, where he has also served as chairman of the department of philosophy and as director of the Center for Philosophy of Science. For over three decades Mr. Rescher has been editor of the *American Philosophical Quarterly*. He was elected to membership in the Institut International de Philosophie in 1970 and was elected member of the Academie Internationale de Philosophie des Sciences in 1984. The author of more than sixty books in various areas of philosophy, works by Mr. Rescher have been translated into German, Spanish, French, Italian, and Japanese. He has lectured at universities in many countries and has occupied visiting posts at various universities in North America and Europe (including Oxford, Konstanz, and Salamanca). He has held fellowships from the J. S. Guggenheim Foundation, the Ford Foundation, and the American Philosophical Society. A former president of the American Philosophical Association (Eastern Division), the C. S. Peirce Society, and the G. W. Leibniz Society of America, Mr. Rescher has also served as member of the board of directors of the International Federation of Philosophical Societies, an organ of UNESCO. One of the few contemporary exponents of philosophical idealism, Mr. Rescher has been active in the rehabilitation of the coherence theory of truth and in the reconstruction of philosophical pragmatism in line with the idealistic tradition. He has pioneered the development of inconsistency-tolerant logics and, in the philosophy of science, the exponential retardation theory of scientific progress based on the epistemological principle that knowledge increases merely with the logarithm of the increase in information. Books about Rescher's work have appeared in English, German, and Italian. His contributions to philosophy have been recognized repeatedly by honorary degrees awarded by universities in the United States and abroad. In 1977 its fellow elected him an honorary member of Corpus Christi College, Oxford, and in 1983 he received an Alexander von Humboldt Humanities Prize, awarded under the auspicies of the Federal Republic of Germany "in recognition of the research accomplishments of humanistic scholars of international distinction."